Resealing of Buildings
A guide to good practice

This Guide is based upon the findings of the RESEAL project which was carried out during the early 1990s by the partnership outlined below.

RESEAL PROJECT PARTNERSHIP

Lead Partner
  Oxford Brookes University
Main Industry Partner
  Taywood Engineering Ltd
Sealant Manufacturer's Consortium
  Adshead Ratcliffe & Co. Ltd
  Evode Ltd
  Fosroc-Expandite Ltd
  Morton International Ltd
  Sika Ltd
Advisors
  Association of Sealant Applicators
  Building Research Establishment
Government Sponsors
  Engineering and Physical Sciences Research Council (EPSRC, formerly SERC)
  Department of Environment (DOE) Construction Sponsorship Directorate

Resealing of Buildings
A guide to good practice

Principal author
Ron Woolman

Edited by
Allan Hutchinson

Butterworth-Heinemann Ltd
Linacre House, Jordan Hill, Oxford OX2 8DP

A member of the Reed Elsevier group

OXFORD LONDON BOSTON
MUNICH NEW DELHI SINGAPORE SYDNEY
TOKYO TORONTO WELLINGTON

First published 1994

© Joining Technology Research Centre, Oxford Brookes University 1994

All rights reserved. No part of this publication
may be reproduced in any material form (including
photocopying or storing in any medium by electronic
means whether or not transiently or incidentally
to some other use of this publication) without the
written permission of the copyright holder except
in accordance with the provisions of the Copyright,
Designs and Patents Act 1988 or under the terms of a
licence issued by the Copyright Licensing Agency Ltd,
90 Tottenham Court Road, London, England WIP 9HE.
Applications for the copyright holder's written permission
to reproduce any part of this publication should be addressed
to the publishers

The views offered in this guide represent a consensus of the
main contributors in the RESEAL project. They do not necessarily
represent individual opinions or reflect advice which may be
given by individual partners.

British Library Cataloguing in Publication Data

Resealing of Buildings: Guide to Good
Practice
  I. Hutchinson, Allan
  690.24

ISBN 0 7506 1859 0

Library of Congress Cataloging in Publication Data

Resealing of buildings: a guide to good practice/edited by Allan
  Hutchinson.
  p.   cm.
  Includes bibliographical references and index.
  ISBN 0 7506 1859 0
  1. Exterior walls–Joints. 2. Sealing (Technology). 3. Calking.
  I. Hutchinson, Allan.
  TH2235.R47
  693'.89–dc20   93–48893
                                                                      CIP

Typeset by TecSet Ltd, Wallington, Surrey
Printed in Great Britain by Redwood Books, Trowbridge, Wiltshire

Contents

Foreword by Dr John Bowler-Reed		xi
Foreword by Professor Sir Edmund Happold		xii
Preface		xiii
Partners in RESEAL project		xv
List of main contributors		xviii
Introduction		xx

1. Movement in buildings – joints — 1
   1.1 Opening remarks — 1
   1.2 Purpose of joints — 1
   1.3 Settlement — 4
   1.4 Moisture — 5
   1.5 Creep — 6
   1.6 Thermal expansion — 6
   1.7 Loading — 11
   1.8 Wind — 11
   1.9 Vibration — 12
   1.10 Measurement of joint movement — 12
   1.11 Summary remarks — 13

2. Consideration of factors in resealing — 15
   2.1 General — 15
   2.2 Suitability of the joint for the amount of movement — 16
      2.2.1 Is the joint wide enough? — 16
      2.2.2 Is the joint deep enough? — 18
   2.3 Suitability of the sealant to accommodate the type and frequency of movement — 18
   2.4 Suitability of the sealant for the conditions of service — 20
      2.4.1 Damage and vandalism — 21
      2.4.2 Will the sealant last? — 21
      2.4.3 Change in conditions of service — 21
   2.5 Other causes for resealing — 22

		2.5.1	Aesthetic considerations : colour and appearance	22
		2.5.2	Blistering	23
		2.5.3	Fire resistance	23
		2.5.4	Staining	24
3	**Examination and inspection**			**26**
	3.1	Identification of causes of failure		26
		3.1.1	Inspection	26
		3.1.2	Visual examination	27
		3.1.3	Sampling	27
		3.1.4	Dimensions	28
		3.1.5	Movement	28
		3.1.6	Sealant identification	28
	3.2	Evaluation		29
		3.2.1	Adhesion failures	29
		3.2.2	Cohesive failures	31
		3.2.3	Surface-cohesive failures	31
		3.2.4	Ageing	33
4	**Assessment of reseal criteria**			**37**
	4.1	Economics		37
		4.1.1	Perspective on the cost of resealing	37
		4.1.2	Access	39
		4.1.3	Resealing options and cleaning techniques	43
		4.1.4	The sealing system	47
	4.2	Selection of the sealant system		49
		4.2.1	Movement considerations	50
		4.2.2	Cleaning and access	54
		4.2.3	Weather and environmental conditions	55
		4.2.4	Check list	57
	4.3	Site trials		57
		4.3.1	Adhesion tests	58
5	**The reseal operation**			**61**
	5.1	Access		61
	5.2	Cleaning and preparation of joint surfaces		61
		5.2.1	Adhesion	61
		5.2.2	Cleaning and preparation	65
	5.3	Cleaning techniques		68
		5.3.1	Bulk sealant removal	68
		5.3.2	Further cleaning	70

		5.3.3 Final cleaning	71
	5.4	Suitability of cleaning method for surfaces	73
		5.4.1 Concrete, stone and brickwork	73
		5.4.2 Metals and coated metals	74
		5.4.3 Timber	75
	5.5	The failed sealant	**75**
		5.5.1 Non-curing sealant residues	76
		5.5.2 Cured sealant residues	77
	5.6	Final preparation	79
	5.7	Priming	79
	5.8	Joint depth	81
		5.8.1 Joint depth control	82
		5.8.2 Back-up materials	84
		5.8.3 Bond-breaker tapes	86
	5.9	Mixing of multi-component sealants	87
		5.9.1 Sealant materials	87
		5.9.2 Mixing techniques	88
	5.10	Gun application	91
	5.11	Tooling	95
	5.12	Masking	98
	5.13	Protection during cure	100
		5.13.1 Chemical attack	101
6	**Quality assurance guidelines**		**102**
	6.1	The survey	102
	6.2	Assessment of joint requirements	103
	6.3	Assessment of sealant suitability	103
	6.4	Site trials	104
	6.5	The reseal specification	104
	6.6	Operator technique	105
	6.7	On-site storage requirements	105
		6.7.1 General	105
		6.7.2 Waste disposal	106
	6.8	Health and Safety (COSHH Regulations)	106
	6.9	On-site quality control	106
		6.9.1 Opening remarks	106
		6.9.2 Joint dimensions	107
		6.9.3 Cleaning	108
		6.9.4 Final preparation and priming	109
		6.9.5 Back-up materials	109
		6.9.6 Sealant application	110

		6.9.7 Tooling	111
		6.9.8 Adhesion	112
7	**Case histories**		**113**
	7.1	Large frequent cyclic movements in thin concrete cladding	114
	7.2	Large cyclic movements in aluminium cladding	115
	7.3	Staining of marble	117
	7.4	Adhesion to oil-contaminated brickwork	119
	7.5	Oversealing gasket glazing in curtain walling	120
	7.6	Blistering acrylic sealant on hardwood	122
	7.7	Oil-based mastic sealant discoloration from contact with damp-proof membrane	124
	7.8	Replacement of hidden seals in curtain walling	124
	7.9	Poor preparation of joints in precast concrete cladding	126
	7.10	Contamination failure following resealing	127
	7.11	Silicone sealant discoloration from contact with pitch-polymer damp-proof membrane	127
	7.12	Extrusion of bituminous material from concrete joints in a swimming pool	129
	7.13	Silicone sealant discoloration from contact with rubber thermal break in window frames	130
	7.14	Sealant blistering caused by water-repellent concrete surface treatment	130
	7.15	Chemical attack of sealant by brick cleaning fluid	132
	7.16	Use of abseil technique	133
	7.17	Smelly sealant	133
	7.18	Staining by silicone sealants	134
8	**Lessons for new build**		**137**
	8.1	The importance of the joint	137
	8.2	The work programme	138
	8.3	Life expectancy	138
		8.3.1 Nature of the sealant	139
		8.3.2 Exposure of the joint to weather	139
		8.3.3 Amount and frequency of movement affecting the seal	140
		8.3.4 Other degrading or damaging influences	140
		8.3.5 Geometry of the sealant joint	141
		8.3.6 Nature of the joint faces and use of primers	142
	8.4	Timing	142

8.5		Resealing	143
8.6		Access	143

Appendix 1 Headings for survey, specification and assessment sheets   144

Appendix 2 Summary of the work completed during the RESEAL project   148

    A2.1   Background   148
    A2.2   Project approach   149
    A2.3   Experimental work   150
       A2.3.1   Overview   150
       A2.3.2   Joint fabrication and curing   151
       A2.3.3   Mechanical testing   152
       A2.3.4   Surface analysis   152
       A2.3.5   State-of-cure studies   153
       A2.3.6   QUV ageing   153
    A2.4   Synopsis of results from the study of sealant performance on contaminated substrates (resealing)   154
    A2.5   Summary resealing guidelines   155

Glossary   157

Information sources, publications and standards   164

Index   167

Foreword by Dr John Bowler-Reed

It is with unadulterated pleasure and enthusiasm that I exhort everyone who has to do with the resealing of joints in building and civil engineering to study this guide. I have not found any evidence of commercial tendencies in it – one could say that it is completely disinterested, except that this might give the wrong impression that it is less than committed to the advancement of successful resealing technology.

It is a truly comprehensive and practical guide to good practice embodying a concentration of the entire period, from its inception in the early 1950s to the present, during which time sealant technology has been developed and refined in the light of experience. Throughout that period the principal author, Ron Woolman, has been really active in this field and it is characteristic of him that contributions of others have been given equal weight in this guide.

It is the failures in weatherproofing of joints, as in most engineering practice, that has led to more knowledge; this guide is primarily devoted to the remedial treatment of those failures. There is also a section, essential for designers, setting out the lessons that have been learned for new construction.

The emphasis throughout is on the practical application of the knowledge of the distinguished contributors. This knowledge embodies the results of a comprehensive programme of testing that was carried out during the RESEAL project, in order to understand and verify the opinions and conclusions of practitioners.

Dr John Bowler-Reed
Consultant

Foreword by Professor Sir Edmund Happold

By and large architects may be good at 'beautifully composing music crystallized in space that elevates our spirits beyond the limitation of time' – but they have not been so successful in keeping out the water above and below.

And even if performance is passed on to a contractor, not all responsibility can be. In law you cannot have authority without responsibility.

At last contractors, suppliers and researchers, with financial help from government, have produced a book on sealing buildings – a guide to good practice. That this book comes through one of the new universities is heartening for the future.

It will be a best seller. It is a brave architect, consulting building engineer or surveyor who has such confidence as not to buy a copy. They now have something to share the blame with!

And very good the book is too.

Professor Sir Edmund Happold
University of Bath

Preface

Sealants play a vital role in joints in maintaining the weather tightness of buildings and engineering structures. There are a variety of reasons however where there is a need to replace the seals in joints, perhaps many times, during the lifetime of a structure. Up until 1992 there were few objective guidelines for the replacement of failed sealants despite this obvious need. This was the more surprising because a far greater volume of sealant materials is used in resealing than for sealing joints for the first time.

The problem of resealing joints in building structures was addressed comprehensively by a DoE/SERC jointly funded project RESEAL in the Construction, Maintenance and Refurbishment LINK Programme. The concept of the three-year project, which ran from 1989 to 1992, developed as a result of detailed discussions with sealant manufacturers, end-users and leading research centres. The collaborative partnership which evolved included: Oxford Brookes University (lead partner), Taywood Engineering Limited (main industry partner), and a consortium of sealant manufacturers, represented by Adshead Ratcliffe & Co Limited, Evode Limited, Fosroc-Expandite Limited, Morton International Limited, Sika Limited.

Advice and direction was also given by: the Association of Sealant Applicators and the Building Research Establishment.

At the beginning of the project a survey was carried out to determine the scope of the resealing problem, the reasons for sealant replacement, and to identify the most commonly encountered sealant and substrate materials. Representative materials were then used in a large programme of experimental and analytical work which was concerned largely with the problem of obtaining satisfactory adhesion to, and therefore adequate joint performance from, a wide variety of clean and contaminated surfaces. In parallel with the underpinning scientific work, a sub-group of individuals concerned themselves with the production of a practice document on resealing buildings. Experimental findings were fed into appropriate sections of the document on surface preparation and cleaning, upon which many of the recommendations are made. *Resealing of Buildings : A guide to good practice* represents the product of that considerable effort and, it is hoped, will be recognized as the industry standard.

The individuals who deserve particular recognition for their role in the production of this guide are identified in the list of main contributors.

Thanks are also due to all members of the Joining Technology Research Centre, headed by Professor A. Beevers, at Oxford Brookes University for their significant contributions, both towards the experimental work of the RESEAL project and the subsequent collation of information for this guide.

Partners in RESEAL project

Joining Technology Research Centre. Joining and fastening technology has been the primary area of research and development in Oxford Brookes University's School of Engineering since the late 1960s. The major commitment is with adhesive bonding and sealant technology, although the facility still offers services in welding and fastener engineering. A large part of the research effort is concerned with adhesion, surface pretreatment and environmental durability aspects of joints for the aerospace, automotive, defence, general engineering, construction and marine industries.

Taywood Engineering Ltd (TEL) is a leading consultant providing a unique blend of design and technology services to the construction industry. TEL's wide range of architectural and design activities is complemented by specialist consultancy and a NAMAS accredited research laboratory. Research is focused on practical and innovative engineering. Sealant and polymer technologies feature prominently in the laboratory's programme to improve the performance of building materials. TEL possesses the second largest full-scale facade testing facility in the world.

Adshead Ratcliffe & Co Ltd manufactures the ARBO range of products which include polysulphide, acrylic and silicone sealants, preformed mastic tapes and many other specialized sealing components including intumescent tapes and sealants used to give enhanced fire protection. Apart from the primary supply to the glazing, building and construction industries, ARBO materials are used for marine and swimming pool applications, caravans, rail carriages and many other fabricated products.

Evode Ltd is part of the Laporte Group. The EVOSTIK range of sealant products supplied to the building and construction industries include acrylics, polysulphides, silicones, oil-based mastics and bituminous materials. A specialist range of sealants has also been developed with particular properties to meet fire protection requirements.

Fosroc Expandite Ltd is part of the Burmah Castrol Chemicals Group specializing in the manufacture and supply of high performance products for the construction industry worldwide. The sealant products include the brand names THIOFLEX and SECOSEAL and utilize all major polymer technologies including polysulphide, polyurethane, epoxy and hot applied materials for applications from sewage treatment works to motorways and factory floors to building facades.

Morton International Ltd, Polymer Systems is part of **Morton International Inc.** (USA). Polymer Systems sells raw materials such as LP liquid polylsulphide polymers (synthetic rubber), solid rubbers, water dispersion and latices, and speciality low temperature plasticizers. Sealants and coatings based on LP liquid polysulphide polymers have found wide acceptance all over the world in industries such as building construction, insulating glazing, marine, automotive and aircraft.

Sika Ltd operates in over 60 countries worldwide and specialize in the development, manufacture and marketing of 'state-of-the-art' construction chemicals. The highly versatile SIKAFLEX range of one-component high performance polyurethane adhesives are widely used in all types of buildings and civil engineering structures and in a wide range of situations. They also have many uses in industrial assembly and installation applications. Other products in Sika's range include acrylics, silicones and fire resistant sealants.

Addresses

Joining Technology Research Centre
Oxford Brookes University
Gipsy Lane Campus
Headington
Oxford OX3 0BP
Tel: 0865 483504
Fax: 0865 483637

Adshead Ratcliffe & Co Ltd
Derby Road
Belper
Derby DE56 1WJ
Tel: 0773 826661
Fax: 0773 821215

Taywood Engineering Ltd
Taywood House
345 Ruislip Road
Southall
Middlesex UB1 2QX
Tel: 081 5754857
Fax: 081 5754044

Morton International Ltd
Polymer Systems
University of Warwick Science Park
Sir William Lyons Road
Coventry CV4 7EZ
Tel: 0203 416632
Fax: 0203 414745

Evode Ltd
Common Road
Stafford ST16 3EH
Tel: 0785 57755
Fax: 0785 41615

Fosroc Expandite Ltd
Pitfield
Kiln Farm
Milton Keynes, Bucks
MK11 3LX
Tel: 0908 261220
Fax: 0908 261467

Sika Ltd
Welwyn Garden City
Herts AL7 1BQ
Tel: 0707 329241
Fax: 0707 329129

Main contributors

Ron Woolman (Principal author)
Ron is now an independent Consultant following his retirement, in 1991, as Technical Services Manager of Adshead Ratcliffe & Co Ltd. He spent 22 years providing technical support and specifications for both new and remedial projects concerned with sealing and glazing, and was responsible for numerous site and laboratory investigations. He has served on many British Standards Institution committees as a representative of the sealant industry, as well as serving on technical committees of the Sealant Manufacturers' Conference (now the British Adhesives and Sealants Association) and the Glass and Glazing Federation. He is the author of many technical reports and recommendations, as well as publications in the open literature, and has assisted in the preparation of various trade publications and standards.

Allan Hutchinson PhD, BSc, CEng, MICE (Editor)
Allan is a Senior Lecturer in the School of Engineering at Oxford Brookes University and Deputy Head of the Joining Technology Research Centre. His teaching and technical interests include all aspects of design, materials engineering and a variety of joining technologies. His research interests lie with adhesive and sealant materials, advanced composite materials, adhesion and surface pretreatment, joint durability and the performance of large-scale bonded structures. He is the author of over 60 technical publications including a monograph on adhesives in civil engineering.

Roger Browne PhD, BSc, CEng, FICE
Roger is the Assistant Managing Director of Taywood Engineering Ltd, the design and consultancy company of the Taylor Woodrow Group. He was involved in establishing their research laboratories in the early 1960s, and his special technical interests include the design life and durability characteristics of construction materials and building structures. He is particularly concerned with the development of realistic laboratory test methods to simulate site conditions, as well as with performance inspection techniques for structures, and this specialization has led to numerous investigations of

deteriorated structures world-wide. He is the author of over 60 technical publications and has served on many national and international committees including, currently, ISO/TC59 dealing with the design life of buildings. He also chaired the committee for the UK Centre for Window and Cladding Technology at Bath which, in 1993, produced the first international standard on curtain walling testing for buildings.

Paul Maton BSc, ARSC
Paul is currently Technical Director of Octavius Hunt Ltd. Between 1987 and 1992 he worked for Ralli Bondite Ltd, a company involved with the manufacture of a wide range of high performance sealants and surface coatings. As Technical Manager he represented the company on technical committees of the British Adhesives and Sealants Association, and the Glass and Glazing Federation. He also served on a number of British Standards Institution committees as a representative of the sealant industry.

Shaun Hurley PhD, BSc
Shaun is Principal Chemist and Manager of the Analytical Group at Taywood Engineering Ltd. His technical interests encompass all aspects of polymer-based materials used in construction, a field where he has over 20 years' world-wide experience. He has published a number of papers and lectured widely in this area, and currently serves on committees concerned with coatings, repair materials, sealants and adhesives.

Introduction

Maintenance and refurbishment now represents a very substantial activity within the building construction industry. Statistics suggest that the volume of maintenance and repair work now equals or exceeds new-build and that the proportion is likely to increase as buildings are repaired and refurbished, adapting them to the needs and standards of modern industry and the modern environment. The market value of construction maintenance and refurbishment in the United Kingdom was estimated to be £17 billion in 1991, or about 50% of the total construction market.

This guide is concerned with joints in buildings. It does not specifically deal with sealed joints involving other structures although the information contained is in fact of wide applicability. A substantial proportion of the cost of building maintenance is attributable to resealing of the external cladding. The modern building sealant industry is a post Second World War development. Prior to 1939, putty and bitumen were the major sealant materials for the construction industry. Changes in methods of construction, together with new materials and innovative design, have increased the volume of sealants used. The increased appreciation of performance requirements has led to the introduction of new materials having improved performance and durability.

The heavily worked external sealant joint in a building is subject to continual motion as well as various physical and chemical degradation mechanisms, and the life expectancy of the sealed joint is usually less than that of the building facade.

The causes, nature, variety and extent of building movements are now becoming appreciated. There is also a better understanding of the way that sealants perform and of their ability to accommodate various forms of building movement, and to remain bonded to a wide variety of building material surfaces. The technology of joint design has also been developed together with a better understanding of sealant application. This has created a technical basis for sealing joints in buildings.

The accumulation of this knowledge has taken many years during which time many buildings have been sealed with varying degrees of success. The industry is therefore faced with a legacy of an increasingly large number of buildings which require resealing, some of which may have inherent defects varying from incorrect sealant or poor application to more serious design faults.

The amount of gunnable sealant used in resealing in the UK in 1990 was estimated at 13,000 tonnes per annum, at an installed cost of at least £500m. This is equivalent to resealing about 100,000 kilometres of joints per year. A modest ten-storey office block may have more than 5 kilometres of joints, whereas a tall building may have more than 50 kilometres of joints and is likely to be even more vulnerable to leakage because of the severity of exposure. The high installation cost reflects the difficulties of the resealing operation. These costs include labour, provision of access to the joints, removal of the old sealant to an adequate standard, priming as necessary, replacement of back-up material, replacement of sealant and tooling to achieve the correct joint profile and appearance.

The resealing process is significantly more complex than sealant installation in new buildings (Figure 1), but many projects are not designed to be resealed despite this obvious need. A large number of joints are placed in inaccessible positions, creating difficulty for the operator and even more difficulty with equipment. In some facades, joints were sealed as the cladding was assembled and may not now be accessible. In such cases it may be necessary to apply ancillary components to form new joints, although in some cases overcladding or recladding may be the only long-term solution.

There is considerable uncertainty regarding the effects of sealant residues or surface coatings on the resealed joint behaviour, and this aspect may require substantial technical investigation and site trials before sealant replacement can be undertaken.

There are many reasons for resealing buildings and, because of the expense and inconvenience involved, it is important to ensure that the resealing operation is successful. The RESEAL project has shown that there are many causes of poor performance and that it is important that these are appreciated if a successful reseal is to be achieved. These causes include:

- Lack of environmental control, and working under unsuitable conditions
- Difficulties of access
- Pressures on completion times
- Lack of appreciation of building movements
- Lack of understanding of joint design principles
- Lack of understanding of sealant technology.

xxii *Introduction*

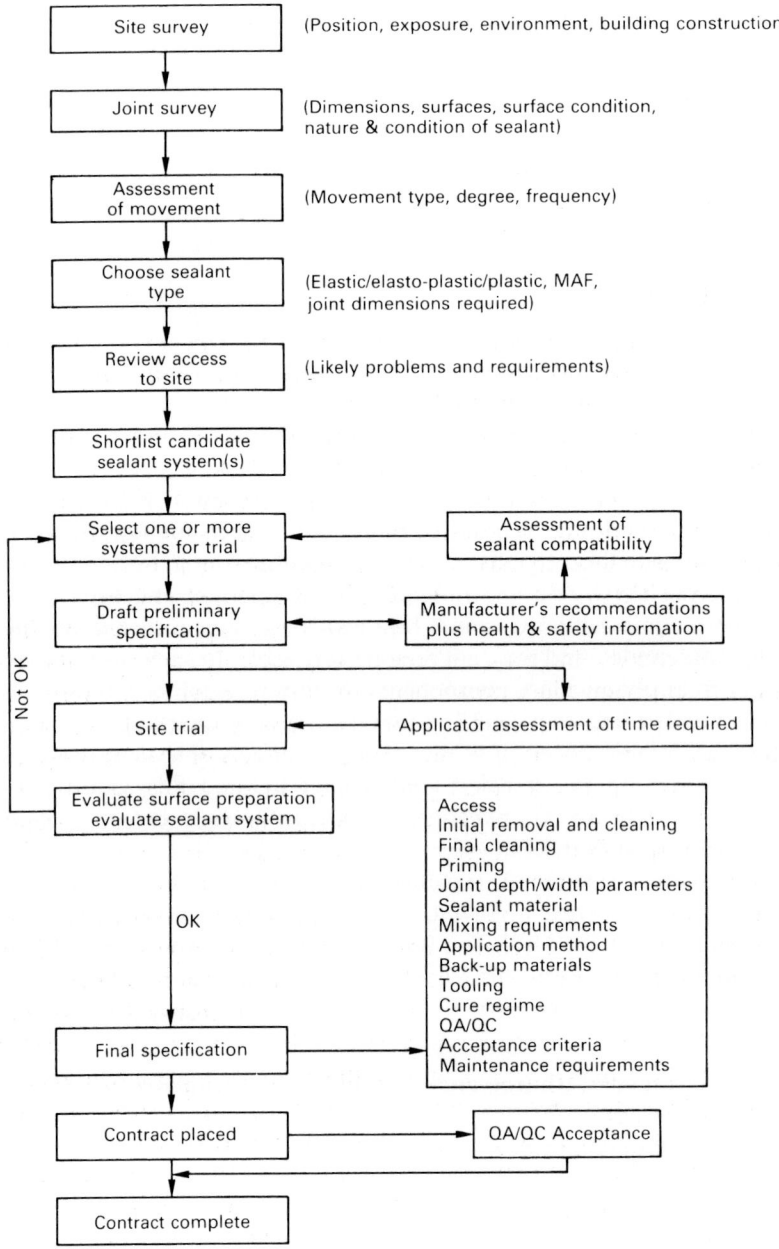

Figure 1 *The reseal process*

This document includes detailed information under eight headings:

1. Movement in buildings – joints
2. Consideration of factors in resealing
3. Examination and inspection
4. Assessment of reseal criteria
5. The reseal operation
6. Quality assurance guidelines
7. Case histories
8. Lessons for new build

There is also an appendix which contains the headings for appropriate Survey, Specification and Assessment sheets. A further appendix summarizes the work carried out during the RESEAL project.

Although the document is primarily concerned with resealing, much of the information is equally applicable to new build. Building movements, in terms of character and amplitude, are primary parameters in selecting sealant types for both new and reseal operations. However the pattern of movement in an old building may be significantly different to that in a newly erected structure. These differences are discussed in this document.

Few sealants are designed specifically for resealing and the sealant properties discussed, in general terms, are equally appropriate for consideration in both new build and resealing. However, there are situations in which sealants used in new build are less suitable for reseal applications as the choice may be influenced by considerations that do not apply to new work. This document provides guidelines to the specifier for the total reseal system, including procedures for site inspection and site trials prior to the final specification, together with check lists and sample sheets in Appendix 1 suitable for controlling work on site.

In parallel with the production of this document, a UK survey was carried out in 1990 with the object of determining the scope of the resealing problem. This led, among other things, to a realization that adhesion failures are a relatively common occurrence. It also emerged that concrete and aluminium were the substrates most commonly encountered in resealing work, and that curing sealant materials were being used in the main for replacement seals – with a strong tendency towards one-part materials, particularly silicones. These findings mirrored those of a Japanese study carried out during the period 1980–1984. The large experimental programme carried out during the course of the RESEAL project was concerned with assessing the performance of newly-sealed and resealed joints, with particular emphasis on adhesion to a wide variety of surfaces and surface conditions. The experimental findings provided a substantial basis for the recommendations and suggestions for surface preparation and cleaning contained in Sections 4 and 5 of this document. It should be noted that the experimental work was carried out on particular sealant/substrate combinations; although the number of experimental combinations was very large the findings should be viewed as specific to the materials used in the investigation. Changes in sealants, primer materials and cleaning solvents may influence the detailed results but should not unduly affect the generic trends found. Appendix 2 contains a brief summary of the experimental work conducted within the RESEAL project.

1 Movement in buildings–joints

1.1 Opening remarks

Movement occurs in all buildings. In traditional forms of building this movement was accommodated by minor cracks and fissures in thick walls or by various design features such as overhangs, flashings, and so on. These techniques involve slow heavy methods of construction and are limiting on design, form and materials. In modern buildings, which are characterized by lighter construction, these movements are accommodated by means of designed joints of various types. Sealant joints are used extensively and when properly designed and sealed, such joints provide a versatile and cost effective solution.

Movement is caused by a variety of effects related to materials, construction, weather and use. Movement takes place in various ways depending on the cause. In order to design joints adequately, it is necessary to appreciate these causes and their effect on the building and the materials used to seal those joints.

Causes of movement include settlement, shrinkage, creep, thermal expansion and contraction, loading, wind and others. Movement induced by a specific cause tends to follow a characteristic pattern but the pattern may be modified by the nature of the construction. It should be noted that many joints are subjected to several types of movement simultaneously.

1.2 Purpose of joints

The joints in a building are discontinuities in the fabric, located in positions between either similar or dissimilar materials and designed to accommodate movements between components or parts of the building. Such movements may be brought about by any of the causes discussed above and may be accommodated in the joint by deforming the seal by compression and extension or by deformation in shear (see Figure 1.1). A number of seal shapes are commonly encountered in joints and these, together with typical failure initiation sites for joints subject to movement, are illustrated in Figure 1.2.

2 Resealing of Buildings

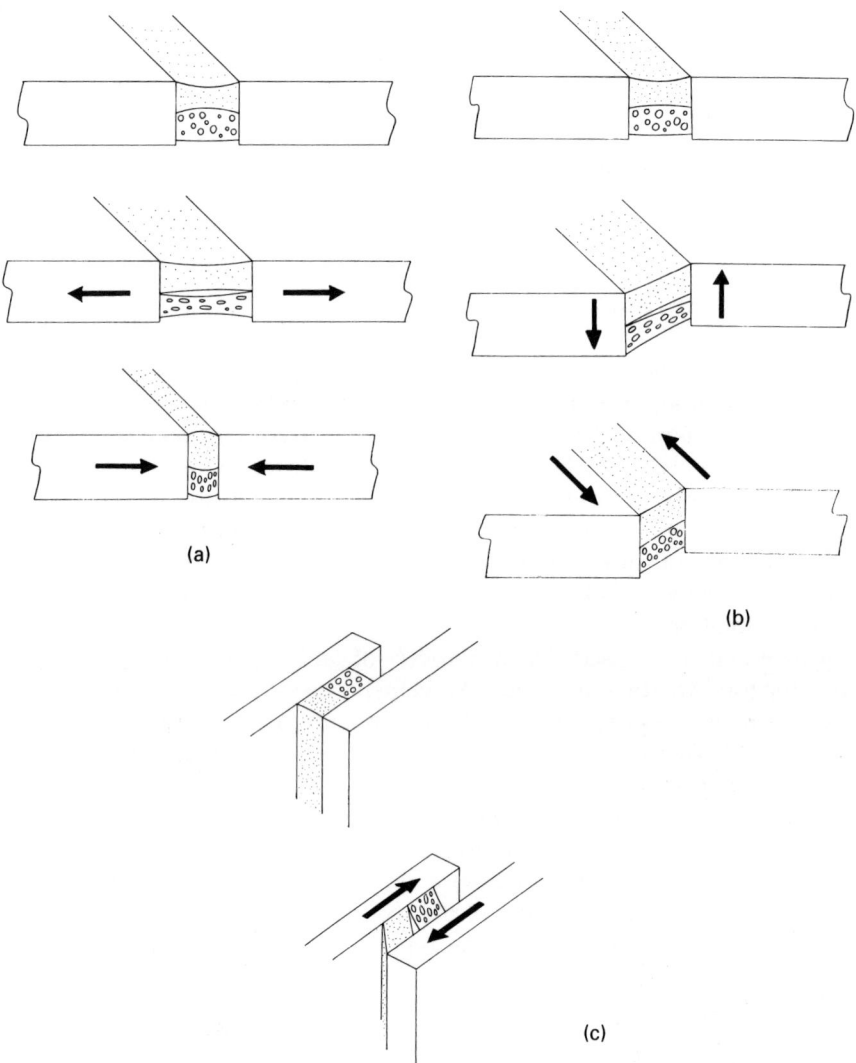

Figure 1.1 *Movement in different types of joint. (a) Butt joint in tension/compression; (b) butt joint in shear; (c) Lap joint in shear*

In order that the seal can deform to accommodate these movements it is essential that the seal should be bonded to the two opposing faces of the joint, leaving the remaining faces of the seal free to deform in order that the

Movement in buildings–joints 3

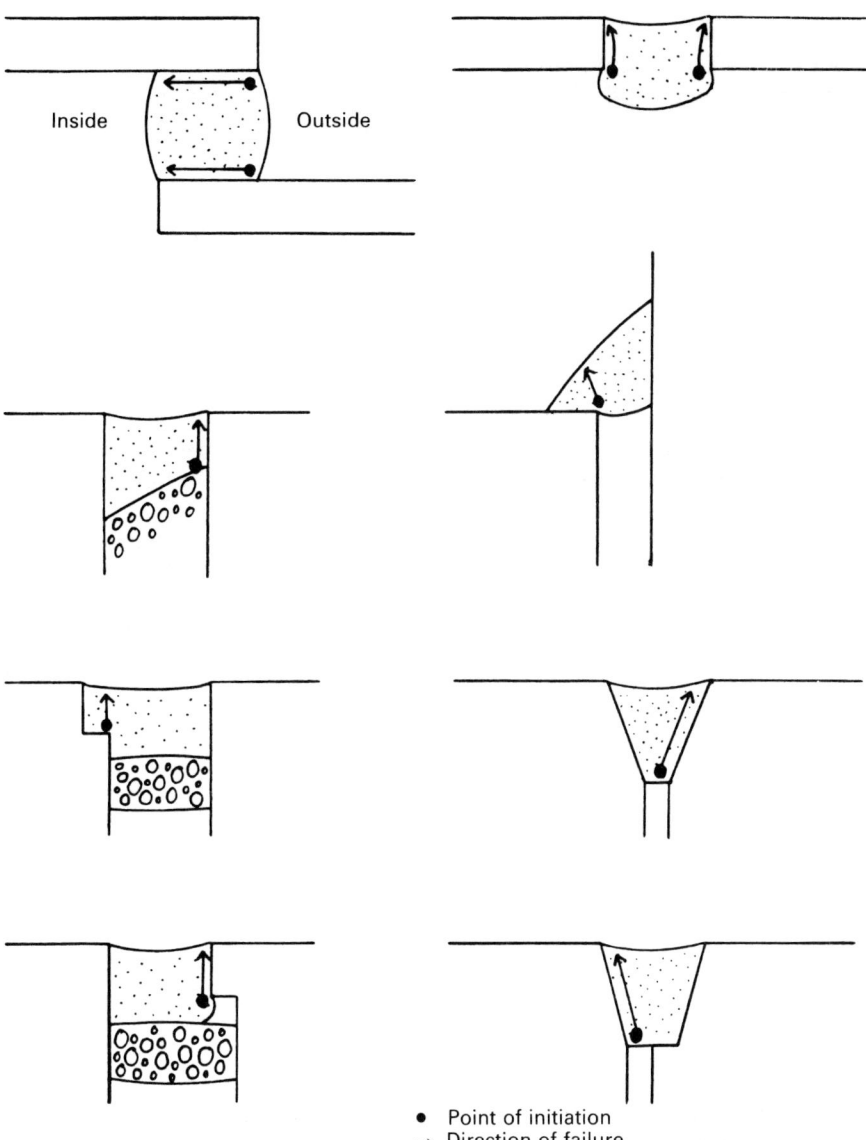

• Point of initiation
→ Direction of failure

Figure 1.2 *Joint geometries, seal shapes and failure initiation sites*

mass of sealant is able to deform without inducing localized stress that would cause premature failure.

Optimum joint depth depends upon the type of movement affecting the joint and the elastic character of the sealant. In joints subject to shear movement, the seal depth (in the direction of movement) should be at least equal to the width. In joints subject to tension and compression movement, the joint depth depends on the elastic nature of the sealant (*see* Section 5.8).

If the joint is deep, a compressible joint filler should be used to fill the inner part of the joint and to control the depth of sealant (*see* Section 5.8). If the compressible filler is a material to which the sealant would stick, it should be covered with a bond-breaker tape to prevent the sealant adhering to the back of the joint and creating third surface (three-sided) adhesion.

The joint filler or back-up assists in the application of the sealant into the joint by controlling the depth of sealant and providing resistance to sealant flow, encouraging the sealant to flow to the sides of the joint to wet the joint faces and to achieve good adhesion.

In the absence of a joint back-up sealant tends to flow through the joint, barely wetting the sides and resulting in poor adhesion. Further, because of the lack of resistance, it is more difficult to tool the sealant to achieve a smooth surface. When sealing joints in thin claddings, such as glass assemblies, it is necessary either to apply a temporary backing to the joint which is removed after cure or, in some cases, it may be possible to apply sealant and tool from both sides simultaneously. This is a difficult specialized technique, requiring considerable skill and practice.

Back-up materials are discussed in Section 5.8.2.

1.3 Settlement

All buildings are affected by settlement as the ground, upon which the building is constructed, deforms to accommodate the loads imposed upon it. The pattern of movement and the time over which it occurs depend on the nature of the land. The extent of deformation may be affected by the size and weight of the structure.

Joints to accommodate settlement movement are placed between new and old structures, between structures of differing size, height or weight, or where ground conditions provide a differing degree of support.

The normal characteristics of settlement are that the movement is slow, primarily in one direction, it takes place over a limited period of time and results in a permanent distortion of the joint. The period of movement is usually several years.

In special cases, where the supporting capacity of the land may vary, settlement can occur at other times. Typical examples are mining subsidence and some areas affected by changes in water table or tidal waters.

1.4 Moisture

Moisture movement occurs in two differing forms. Some building materials, such as concrete, brickwork and blockwork, exhibit setting or drying out shrinkage resulting in a permanent change whereas others, such as timber, vary in size with change in moisture content resulting in a cyclic pattern of movement.

The structural framework and floors of many larger buildings are formed from concrete, brick or blocks using large quantities of water which result in the structure having a high moisture content. This moisture dries out as the building comes to completion and during the early life of the project. This drying out causes the structure to shrink and, if joints are not provided in appropriate places, this results in cracking and consequential problems.

Where joints are placed in suitable positions, the joints tend to open as the structural sections dry out and shrink. Movement is generally slow, primarily in one direction and results in permanent distortion of the joints.

Where materials having differing shrinkage or drying rates are used in combination, for instance in-situ concrete and brickwork, the rate and extent of movement will vary and this may result in a closing or shearing action at some joints.

Many materials are prone to expand when wet and shrink when dry. This is obvious with materials such as timber but also occurs in materials such as concrete, brick, stone and many others although to a lesser extent. When used internally, such materials tend to dry to a low moisture content and remain at a relatively stable condition resulting in a permanent shrinkage, for example, shrinkage of joists, skirtings, architraves, and so on.

When used externally these materials are subject to the vagaries of the weather and their moisture content may vary widely and frequently. The moisture content of stained timber has been measured at over 30%, that is saturation, after prolonged rain whereas the moisture content after a dry spell in summer fell to 6%. Considerable changes in moisture content can occur in a few hours, especially with dark coloured finishes. Such changes can cause considerable changes in dimensions, especially cross-grain in timber. In this case the resultant movement is relatively rapid, repetitive and cyclic in nature.

An indication of the reversible moisture movement of building materials is given in Table 1.1.

Table 1.1 Reversible moisture movement of building materials

Material	Expansion Dry to saturated (%)	Over 3 m length (mm)
Dense concrete	0.02–0.06	0.6–1.8
Light aggregate concrete	0.03–0.06	0.9–1.8
Ultra-lightweight concrete	0.1–0.2	3.0–6.0
Fibre/cement board	0.1–0.25	3.0–7.5
Limestone (average)	0.01	0.3
Sandstone	0.05–0.08	1.5–2.4
Brick (clay)	0.003–0.02	0.09–0.6
Brick (sand–lime)	0.01–0.05	0.3–1.5
Glass reinforced cement	0.15–0.25	4.5–7.5
Timber (with grain)	0.05–0.1	1.5–3.0
Timber (across grain)	2.0–8.0	Not applicable

1.5 Creep

Creep is a phenomenon primarily associated with concrete-framed structures. Concrete shrinks as it sets and dries, but experience shows that it continues to deform when under load, long after setting and drying are complete. Tall buildings get gradually shorter and the distance between floor and ceiling gets less. Beams continue to deflect and sag. The effect is gradual, taking place over a period of years, and movement is believed to continue for at least 30 years although in most cases the major portion takes place in 15 years. Additional creep may however be induced by a change of use.

The amount of movement is small, probably not more than 3–4 mm per floor height, but can cause serious damage if not taken into account in the design of the project. The British Standard Code of Practice for Brick and Concrete Cladding requires a 13 mm compression joint at each floor level on concrete framed buildings to accommodate this effect.

Movement due to creep is very slow, it is load-directional, compressive and takes place over many years causing permanent distortion of the joint.

1.6 Thermal expansion

Thermal expansion is a more complex form of movement. Most materials expand when they get hot and shrink as they are cooled. The extent to which they expand or contract is a characteristic of the material known as its coefficient of linear thermal expansion and is specific to that material (Table 1.2).

Materials such as glass, brick, steel, stone and concrete have relatively low coefficients of linear thermal expansion whereas that of aluminium is two times greater and most plastics materials have very high coefficients of

Table 1.2 Typical linear thermal expansion of building materials

Material	Coefficient of expansion per °C × 10⁻⁶	Expansion of 3-metre length over temperature differential of: 70°C (mm)	85°C (mm)	100°C (mm)
Clay bricks	5.0	1.05	1.3	1.5
Concrete (light aggregate)	8.3	1.75	2.1	2.5
Concrete (gravel aggregate)	11.7	2.45	3.0	3.5
Concrete (limestone aggregate)	6.0	1.26	1.53	1.8
Aluminium	23.5	5.0	6.1	7.15
Stainless steel	18.0	3.78	4.6	5.4
Structural steel	12.1	2.54	3.1	3.63
Copper	18.0	3.8	4.6	5.4
Bronze	19.8	4.16	5.05	5.94
Glass	9.1	1.91	2.32	2.73
Limestone	2.5–9.0	0.5–1.9	0.65–2.3	0.75–2.7
Granite	8.5–11.0	1.8–2.3	2.1–2.8	2.5–3.3
Sandstone	7.0–16.0	1.5–3.3	1.8–4.1	2.1–4.8
Marble	13.2	2.77	3.37	4.0
Timber (with grain)	3.8–6.5	0.8–1.4	1.0–1.67	1.14–1.95
Timber (across grain)	50–60	10.5–12.6	12.7–15.3	15.0–18.0
Glass reinforced cement	7.0–12.0	1.5–2.5	1.8–3.1	2.1–3.6
Acrylic sheet	70–90	14.7–18.9	17.9–23.0	21.0–27.0
Polyester GRP	18–50	3.8–10.5	4.6–12.7	5.4–15.0
Vinyl sheet (PVC)	40–75	8.4–15.8	10.2–19.2	12.0–22.5

70°C differential is appropriate for *light* coloured *uninsulated* or *heavy* cladding.
85°C differential is appropriate for *light* coloured *insulated* or *dark* coloured *uninsulated* cladding.
100°C differential is appropriate for *dark* coloured *insulated* cladding.

Note: Even greater temperature differentials may occur with high standards of insulation used with high absorption finishes.

expansion. Some materials, including stone and timber, have different coefficients of expansion in different directions.

The coefficient of thermal expansion is the amount a material expands or contracts for a temperature change of one degree, but the amount the component expands or contracts depends upon its size, its coefficient of expansion and the change in temperature of that component (see Table 1.2). The primary source of heat affecting the exterior of a building is the sun which heats the external surface. The temperature reached by the cladding depends upon the amount of heat energy and what happens to it after it has been absorbed by the cladding.

Materials such as stone and concrete are dense and heavy and require a large input of energy to raise their temperature. Such components react very slowly to sunshine resulting in a slow rise in temperature and a very slow rate of expansion. The daily cycle of temperature change is therefore relatively small compared to the annual cycle, winter to summer. Consequently the daily cycle of movement is small compared to the annual cycle, producing modest fluctuations on the annual cycle. Joints are therefore subject to prolonged but not permanent displacement.

Thinner, lighter cladding materials such as aluminium, plastics, and so on require only a small amount of energy to raise their temperature and hence they react quickly to the heat of the sun. If the cladding is insulated from the main structure and the interior of the building, it cannot dissipate that heat and the cladding can get very hot, reaching temperatures far above ambient.

Similarly, in cold weather, the cladding cools quickly and the insulation prevents the leakage of heat from the building so that the cladding becomes as cold as the outside temperature, and in some cases colder.

Figure 1.3 depicts some typical movement profiles of materials, and Figure 1.4 illustrates the difference between daily and diurnal temperature cycles.

Black aluminium cladding on a well insulated building in the north-west of England has been shown to vary from 15–20°C overnight to over 80°C during the day in summer. In winter, temperatures fall to −20°C overnight but rise to 50°C and more during the day. The annual cycle of temperature change is −20°C to +85°C, i.e. 105°C, whereas the daily cycle may be 60–70°C. The cladding on such buildings undergoes frequent, large cycles of temperature change accompanied by cycles of expansion and contraction. Joints in the cladding are subject to large frequent fast cycles of expansion and contraction.

Many of the light cladding materials have high coefficients of thermal expansion, for example aluminium, PVC, GRP, polycarbonates, etc. Large

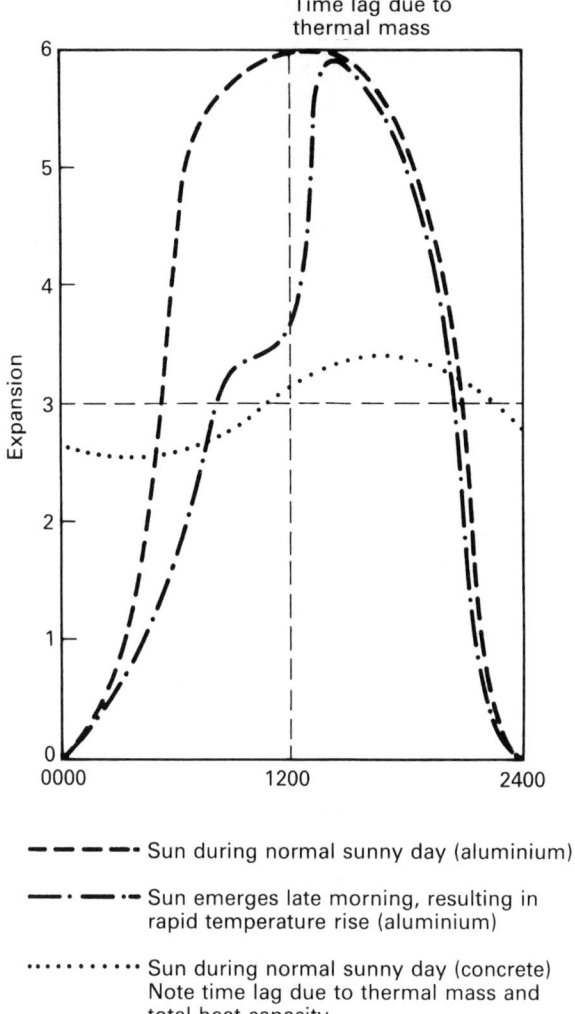

– – – – Sun during normal sunny day (aluminium)

–·–·– Sun emerges late morning, resulting in rapid temperature rise (aluminium)

·········· Sun during normal sunny day (concrete) Note time lag due to thermal mass and total heat capacity

Figure 1.3 *Typical movement of a 3-m length of material over a 70°C temperature change*

changes in temperature are accompanied by large changes in dimension and joints are subject to large, frequent cycles of movement.

The colour and the finish also have a marked effect on the temperature achieved. Everyone is aware that black motor cars get hotter than white cars on a sunny day. The same applies to cladding materials. White components

10 Resealing of Buildings

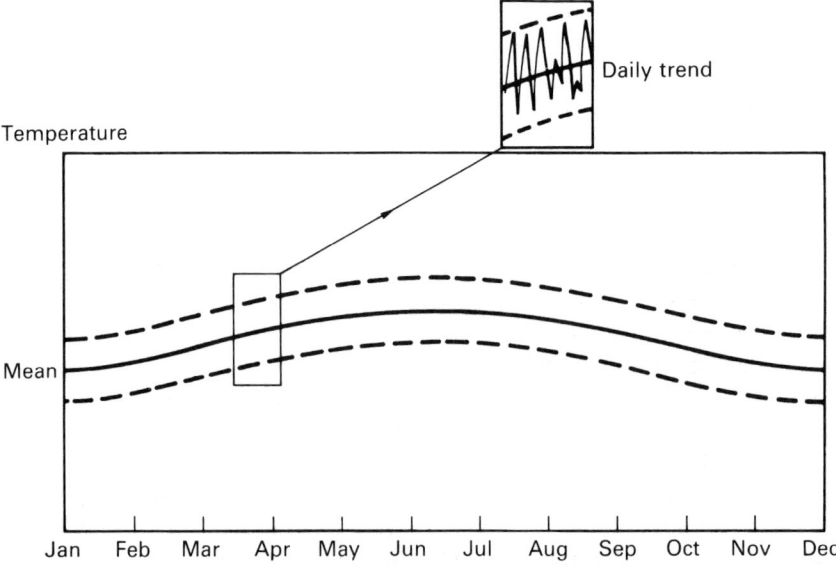

Figure 1.4 *Temperature variations: diurnal trend with daily fluctuations superimposed*

seldom achieve temperatures greater than 50°C whereas matt black components frequently achieve temperatures in excess of 80°C. Shiny gloss finishes absorb less heat than matt finishes. Dark colours absorb more heat than light colours. As a result light-coloured shiny cladding suffers less movement than similar dark-coloured matt finish cladding. These factors should be considered when changing colour schemes. Changing to darker colours may increase temperature changes and hence increase thermal movement putting joints at risk. Changing to lighter colours is a safer option.

Insulation affects all forms of cladding, including walls and roofs. The amount of insulation controls the dissipation of heat from the cladding to the building in summer and the transfer of heat from building to cladding in winter. Light cladding systems that are poorly insulated tend to cushion the effects of weather changes by dissipating heat from the cladding to the building in hot weather, and by gaining heat from the building in winter. This slows the rate of change and reduces the extremes so that the cladding experiences a reduced degree of movement and slower movement. However, the cladding on a well insulated building suffers a wider variation in temperature as the cushioning effect is reduced by the insulation. This factor should be considered in carrying out renovations or improvements.

In carrying out upgrading of industrial units, two-inch thick fibreglass roof insulation was replaced by five-inch insulation. The following summer the aluminium roof got hotter than before, the fixings restricted longitudinal expansion and the roofing sheets buckled, splitting the side seams and tearing out pop rivets. This caused catastrophic leakage in summer storms.

Thermal expansion is a complex subject. BS 8200, British Standard Code of Practice for Design of Non-loadbearing External Vertical Enclosures of Buildings, provides guidance on assessing the various factors. Movement due to thermal expansion has the common characteristic that it is cyclic and repetitive. However, the nature of the cycles may vary.

With heavy cladding materials the annual cycle of movement is large relative to the daily cycle. As a result the joints will be held in a deformed state for several months placing a prolonged strain on the seal. This situation can be relieved by using a sealant with plastic, stress-relieving properties. However, the joint is also subject to cyclic movement and hence requires elastic characteristics. Elasto-plastic sealants are therefore the most suitable.

Aluminium and plastic claddings exhibit larger movements but the daily cycle is frequently large with the seal rapidly varying between tension and compression. The seal is not likely to be stressed for long periods, and it is more important that it should be elastic to accept the cyclic movement.

1.7 Loading

Many components deform when placed under load and recover when the load is removed. Some components in a building are under permanent load and therefore subject to permanent deformation. In other cases components are subject to variable or transient loads and the degree of deformation varies with the loading. This produces a cyclic effect but the periodicity of the cycle depends on the frequency of load change. Joints in bridges subject to large fast changes in load require elastic seals, whereas joints around floors carrying semi-permanent heavy equipment are better sealed with elasto-plastic materials.

1.8 Wind

Wind is a variable factor both in intensity and direction. By its nature it is generally gusting, particularly at the higher speeds and hence places a short term load, pressure or suction on the component or structure. Movement due to wind tends therefore to be cyclic and of short periodicity. Most sealant materials react elastically to short cyclic stress but as the cycles are extended the more elastic materials perform better. Although most sealants perform adequately at low level, the more elastic sealants are more suitable for exposed or high buildings where prolonged winds are more common.

Information on wind speeds experienced in different parts of the UK, together with factors relating to building size, aspect and orientation, are given in British Standard Code of Practice CP3:V:Part 2.

1.9 Vibration

The movement of many joints in buildings is restrained by friction between components, and other factors such as corrosion products, grit, etc. in joints prevents the free movement, expansion or contraction of the components.

Vibration is a form of movement in its own right but it may have a greater effect in relieving friction between components and the consequential lubricating effect, removing the restraint on movement and allowing greater movements that would normally occur in such construction. Vibration may permit the sudden relief of pent-up stresses causing sudden, large movements in structures that normally exhibit small slow movements.

In some cases the vibration may emanate from internal sources such as heavy machinery but in others the vibration may be due to external influences such as traffic.

Vibration has a liberating effect, reducing friction and allowing the unrestrained movement between components whereas friction and other restraints normally inhibit movement to some degree.

Such effects may occur after the building has been substantially completed. The installation of manufacturing plant or the approach of frequent heavy traffic when the building goes into service may cause vibration and consequential failure of joints that had previously been sound.

1.10 Measurement of joint movement

One advantage in resealing buildings is the existence of the building, which can be examined and studied, whereas at the design stage it is a conception and movement can only be calculated or estimated. Joint width can be designed but the reality is always susceptible to the tolerances of the contractor.

It is possible to carry out tests and to collect data on an existing building, prior to resealing, and this may be especially valuable if joint failure has occurred due to movement. There are various means of measuring the movement of joints including the use of scratch tell-tales and transducers which can be attached to the building to obtain information. It should however be appreciated that such devices may need to be observed over a long period to allow for the variety of weather conditions that may affect the building. It may also be necessary to instrument a large number of joints in order to allow for variations in joint movement and ease of joint movement and exposure although some guidance may be obtained from the pattern of joint failure.

Instrumentation may be especially useful where joints appear to move excessively as knowledge of the degree and nature of movement may assist in determining the cause and consequential likely amplitude etc. which will assist in designing the reseal system.

It is important that weather conditions should also be recorded throughout the period of joint observations in order that changes that occur during the observation period can be related to extreme conditions. It should, however, be appreciated that friction and other movement-inhibiting effects may have a greater influence over a limited range of conditions than over the extreme range. Consequently the wider the range of weather conditions and the longer the period of observation, the more useful the results.

1.11 Summary remarks

There are other causes of movement in buildings. These can be analysed and the pattern and extent of movement assessed in order to choose the most appropriate sealant system, and to design a suitable joint.

Many joints are affected by more than one type of movement. In many cases the so-called main expansion joint has to accommodate settlement, drying-out shrinkage and thermal expansion. In choosing the sealant, consideration should be given to the nature and extent of the individual movements and the cumulative effect. Sealants having both elastic and plastic properties are frequently required. Having chosen the sealant, the joint can be designed taking into account the extent and direction of the various movements.

In resealing joints, it may be found that some of the permanent distortion effects have already taken place and their effect is therefore less critical. It may be possible to select a better seal to suit the less onerous conditions. For instance, most concrete and brickwork shrinkage will have taken place within 3 to 5 years. The majority of creep occurs within the first 15 years. Most of the settlement effects occur within 5 years on normal land unless it is affected by changes in the water table. It may be possible to ignore such movements when considering materials for resealing.

14 Resealing of Buildings

For similar reasons it is always advisable to leave the sealing of structural joints in new buildings as late as possible to allow as much of the permanent deformation to take place before sealing and thus minimize the stress on the joints, reducing the risk of failure.

2 Consideration of factors in resealing

2.1 General

It may seem obvious that a building needs resealing when it leaks, but it should be appreciated that leakage is a sign that failure has already occurred and damage has already been done. It is advisable to carry out regular inspection of joints and, where possible, *to replace seals before complete failure in order to prevent consequential damage.* Water penetration may be due to seal failure but could also be due to other causes. There is always a need to identify the source of water penetration, which may not be directly attributable to seal failure, as well as to identify seal failure and its cause before embarking on a programme of resealing. Failure to identify the cause properly may result in a repeated problem and further consequential damage.

Sealant failure can occur for a variety of reasons. The following questions should be considered:

- Were the joint and sealant suitable to accommodate the amount of movement occurring?
- Was the sealant suitable to accommodate the type and frequency of movement?
- Was the seal suitable to accept the conditions of service, e.g. load, pressure, traffic, etc?
- Was the seal suitable to accept damage and/or vandalism?
- What was the service life of the sealant material?

Replacement of the original seal with the same material may not be suitable because this may perpetuate the problem.

> *A twenty-storey tower, originally sealed with an acrylic sealant, was resealed four times with the same material in twenty years but began to leak again within a few months each time. It was eventually cleaned and resealed with a silicone sealant, a difficult and expensive operation, but worthwhile as the building has since been weathertight for more than ten years.*

'Horses for courses' is very true of sealants. The acrylic sealant was not able to accommodate the form of movement or the amplitude of movement occurring in the light aluminium curtain wall of the building. It was used for resealing because of the difficulty of removing the old sealant.

2.2 Suitability of the joint for the amount of movement

There are only two types of joint to accommodate movement – butt joints and lap joints. Butt joints accept movement by deforming in extension and compression. Lap joints accommodate movement by deforming in shear.

Lap joints can accept larger movements than butt joints but are generally more difficult to incorporate and to seal; they can also be very difficult to reseal.

2.2.1 Is the joint wide enough?

The various sealant materials have very different Movement Accommodation Factors – that is their ability to accommodate movement without failure – as defined in British Standards BS 6093 and BS 6213. The Movement Accommodation Factor (MAF) is the amount of movement a butt joint can accommodate, expressed as a percentage of the minimum joint width (see Table 2.1). Movement is considered as total joint movement, that is the change in dimension from maximum to minimum. There is no specific test for the Movement Accommodation Factor. It is assessed by the sealant manufacturer based on a variety of test results and experience of the behaviour of sealants in various forms of construction.

In order to ensure that a joint is adequate to accommodate the movement regardless of its condition at the time of sealing, design joint width is calculated by the following formula:

Design joint width = (Movement × 100 ÷ MAF) + Movement

For example: Joint movement 5 mm, Sealant A has MAF = 50%.
Joint width required = (5 × 100 ÷ 50) + 5 = 15 mm.

For sealant B having MAF = 25%
Joint width = (5 × 100 ÷ 25) + 5 = 25 mm.

Sealant B requires a joint at least 25 mm wide whereas sealant A only requires a joint 15 mm wide.

Table 2.1 Typical Properties of Sealants

Sealant type	Movement Accommodation Factor (%)	Character	Life expectancy (years)	Joint suitability
Oil-based	10	Plastic	1–10	Perimeter pointing.
Butyl-based	10	Plastic	15–20	Concealed joints. (not UV resistant.)
Acrylic				
Water-based	15	Plastic	10–15	Internal joints, plaster cracks, etc.
Solvent-based	20	Plastic	15–20	Perimeter pointing, concrete, stone cladding, etc.
Polysulphide				
One-part	20–25	Elasto-plastic and elastic	20–25	Perimeter pointing, structural joints, stone cladding, etc.
Two-part	25–30	Elasto-plastic and elastic	20–25	Structural joints, stone and cladding, joints subject to early high movement.
Two-part high modulus	10–20	Elasto-plastic and elastic	20–25	Paving, traffic, floor joints, etc.
Polyurethane				
One-part	10–30	Elastic and elasto-plastic	20–25	Light cladding, curtain walling, structural joints, stone cladding etc.
Two-part	20–30	Elastic	20–25	Light cladding, curtain walling, paving, etc.
Silicone				
Low modulus	50–70	Elastic and elasto-plastic	25–30	Perimeter pointing, curtain walling, stone and concrete cladding, structural joints, etc.
High modulus	20–30	Elastic	25–30	Glazing, sanitary ware, etc.
Flexible epoxy	5–15	Elasto-plastic	10–20	Floor joints, traffic areas, etc.

2.2.2 Is the joint deep enough?

Joint depth is a second factor because the joint must be deep enough to accept the correct depth of sealant and a suitable back-up or bond-breaker. The depth of sealant depends on the type and the conditions of service.

The elastic characteristics as discussed in Section 2.3 affect the preferred section of sealant to achieve optimum performance. Plastic sealants should be applied as deep as they are wide to achieve optimum performance. Elasto-plastic sealants are applied such that depth is half the width for maximum movement accommodation. Elastic Sealants perform best in thin sections and are usually applied 6–12 mm deep.

Sealants are frequently used in deeper sections where they are subject to pressure, loading or traffic, such as pedestrian walkways (Figure 2.1). This reduces the MAF slightly. *Optimum performance is obtained when sealants are supported by a compressible back-up such as polyethylene foam to which the sealant does not adhere.* This allows the sealant to release from the backing so that the seal is free to extend and compress without impediment. The back-up foam should be at least 20% wider than the joint to ensure that it is compressed adequately to follow the joint movement, and the depth should be at least half the uncompressed width to prevent buckling in the joint. High movement joints may require an even greater degree of back-up compression, depending upon the conditions at the time of resealing. Where rod-form back-up is used it should be at least 30% wider than the joint. All polyethylene foam backing should have a surface skin or a fine texture closed cell structure to minimize sealant adhesion.

2.3 Suitability of the sealant to accommodate the type and frequency of movement

Sealants are applied as liquids or pastes, firming or setting in place to form a solid mass with elastic and/or plastic properties. Setting or firming may be caused by evaporation of solvents, chemical cure, oxidation and polymerization of drying oils, cooling of hot melts, or by other means. Sealants can be separated into three general types based upon their elastic characteristics.

Elastic sealants

Elastic sealants deform reversibly under applied load and are best suited to joints affected by rapid cyclic movements. If held in a distorted state, they are

Consideration of factors in resealing 19

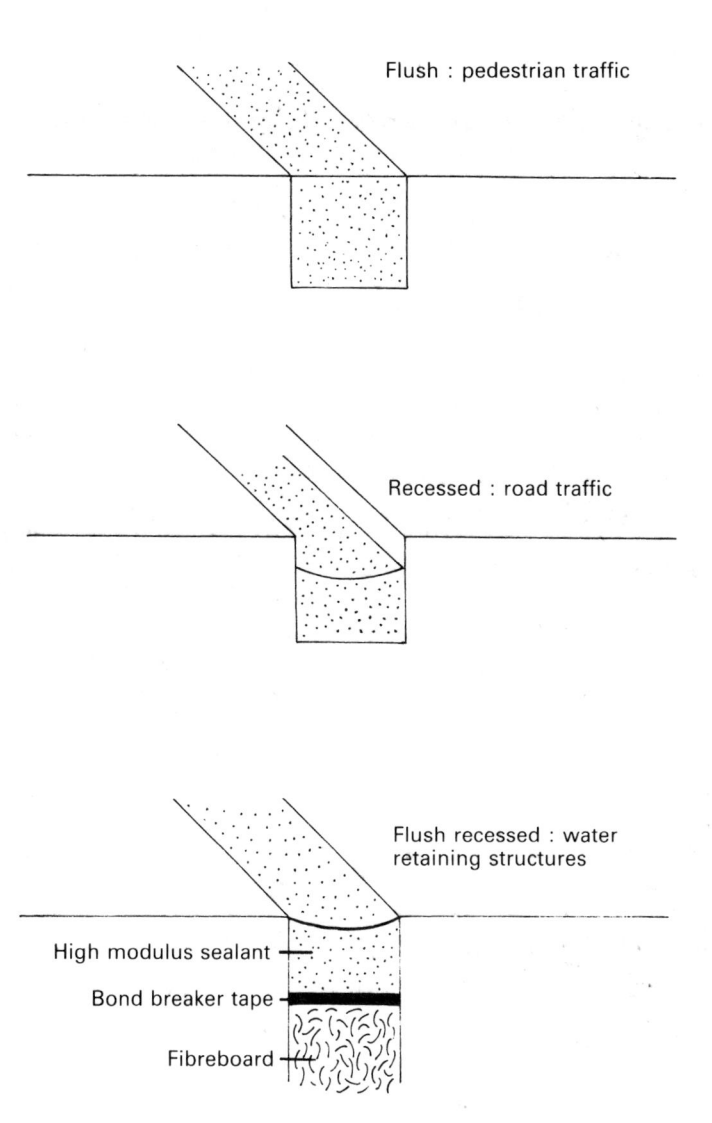

Flush : pedestrian traffic

Recessed : road traffic

Flush recessed : water retaining structures

High modulus sealant
Bond breaker tape
Fibreboard

Figure 2.1 *Recommended joint details for different applications*

subject to prolonged stress which increases the rate of degradation, increases the probability of adhesive failure and increases the risk of tearing or splitting from minor defects. Elastic sealants, such as silicones, are best suited to *curtain walling and insulated light cladding systems.*

Plastic sealants

Plastic sealants do not deform reversibly under an applied load; they stress-relax when held in a deformed state and thus minimize the risk of adhesion failure. However, if they are subject to large, frequent cycles of movement they crease, fold and eventually split. Plastic sealants, such as GP mastic, are best suited to *low movement joints* in structures and other joints subject to *prolonged deformation*.

Elasto-plastic sealants

Elasto-plastic sealants possess both elastic and plastic properties to a reduced degree, partially deforming reversibly under an applied load. Different sealants have differing properties, varying from acrylics, which are primarily plastic with slight elastic character, through the polysulphides, to some low modulus silicones which are primarily elastic but if held in a stressed condition gradually relax over a week or two. The stress relaxing feature reduces the risk of adhesive failure or degradation and the more elastic sealants are better able to accommodate cyclic movements. Elasto-plastic sealants are particularly suited to joints affected by slow prolonged cycles of movement and joints subject to more than one form of movement. *Structural joints, precast concrete cladding, stone cladding and brickwork joints* are typical applications.

The character of various generic types of sealants is given in Table 2.1.

2.4 Suitability of the sealant for the conditions of service

If there are no loading or other factors, low modulus sealants are to be preferred as they impose less strain on adhesion and on the substrates, reducing the risk of failure. However, if the joint is subject to pressure a tougher sealant is generally required. Where joints are subject to prolonged pressure, such as in water retaining structures, a firm support to the seal is necessary in addition to a firm seal – usually applied at least as deep as it is wide (Figure 2.1).

In joints subject to temporary pressures such as traffic, the deeper seal using a tougher sealant is usually adequate applied over a foam back-up. In joints subject to road traffic it is normal to recess the seal to reduce the risk of damage, but this is not desirable in pedestrian traffic areas. In trafficked joints where it is necessary to use a softer seal to accommodate movement, or if joints are excessively wide, it is advisable to use cover plates to protect the seal or to use a mechanical jointing system.

2.4.1 Damage and vandalism

It is generally not possible to use vandal-resistant sealants in joints which are subject to substantial movement. Sealants which resist vandal attack tend to be much tougher, higher modulus and hard materials, generally epoxy resin-based products, having a very low Movement Accommodation Factor. If used in joints affected by movement they are liable to cause damage to the components forming the joint. Where it is necessary to use a soft sealant in areas subject to vandals, consideration should be given to recessing the joint and protecting it with a cover plate or a suitable decorative trim, or incorporating a secondary sacrificial seal to protect the primary seal.

2.4.2 Will the sealant last?

The life expectancy of sealants will vary from as little as a year to thirty years or more depending on type, surface preparation and how it is applied.

Life expectancy is defined as the average period over which the sealant is expected to perform satisfactorily under normal weather conditions, providing it is applied in accordance with the manufacturer's recommendations and instructions. Life expectancy is an average and sealants which are subject to more severe conditions, such as strong hot sun and severe rain, may tend to have a shorter life whereas those in protected or sheltered situations will last longer. Regular inspection is therefore advisable to minimize the risk of consequential damage due to water penetration following joint failure.

Guidance to life expectancy is given in Table 2.1.

2.4.3 Change in conditions of service

Changes to components or change of use may affect the performance of joints and can cause premature joint failure. Change of colour or finish may affect the degree of movement affecting a joint due to change in solar gain. This is particularly true where light coloured components are repainted in darker colours during refurbishment. The colour change may not be intentional.

> *In refurbishing a multi-storey car park, the white concrete top deck that had worn smooth and slippery, was coated with carborundum grit embedded in a black epoxy adhesive. The expansion joints in the deck, that appeared sound when the work was done, failed within a few months because of the increased movement resulting from the increased heat absorption by the black coating.*

2.5 Other causes for resealing

There are some occasions when it becomes necessary to repair or replace the seal although it is technically sound. Some such instances and procedures are described below.

2.5.1 Aesthetic considerations: colour and appearance

Sometimes it becomes apparent on completion that the appearance of the finished joint is not compatible with the designer's concept, either because the colour is not acceptable or because of some other feature. It may be that a recessed joint becomes desirable or unacceptable in order to achieve the desired appearance.

In some cases it may be possible to achieve the desired effect by overcoating the existing seal. Provided that the sealant has good intercoat adhesion, that is new sealant will stick to old, a 2 mm layer of new sealant will usually achieve a change in colour. Most sealants have good intercoat adhesion but some polyurethanes and silicones show poor adhesion of new sealant to old material. With some sealants the older the initial seal, the poorer the intercoat bond. If the adhesion is poor, complete separation may occur as the sealant ages. Solvent cleaning or priming does not always improve the bond between new and old.

Where new sealant will adhere to old, a recessed joint can be made flush. The increased depth of sealant will have only a marginal effect on seal performance unless the recess is very deep. Where new sealant does not bond satisfactorily to old, it becomes necessary to remove the old sealant completely and reseal the joint, usually replacing back-up and primer. It may be necessary to replace a flush joint with a recessed seal. This may be necessary where the designer wishes to create a shadow line. Recessed joints are also desirable where joints may be subject to appreciable movement causing temporary or permanent displacement (Section 2.2.2).

Substantial movement during cure can cause permanent distortion of the seal, causing uneven bulges and hollows which contrast badly with smooth cladding either side. It is not possible to prevent movement during the cure period although careful choice of season to carry out the work may reduce the effect.

Recessing the joint frequently reduces the movement affecting the seal, reducing the distortion, but it also removes the comparison of the smooth surface of the panels with the undulating surface of the sealant giving a more pleasing appearance.

Remedial work may involve the complete removal of the seal and replacement in a more suitable position.

2.5.2 Blistering

Blistering of the sealant may occur soon after application, before the sealant has cured, producing an uneven and untidy appearance. Such blistering may be due to several causes, but it is not due to air or bubbles entrapped in the sealant as applied. Air bubbles trapped in the gun or cartridge are compressed to about one quarter of their original size by the gunning pressure with the result that they expand violently as they pass through the nozzle and burst with a violent popping noise. Any small bubbles that may remain in the seal have no significant effect. A temperature rise of 40°C increases the diameter of a bubble by less than 5%, which would not be visible. Blisters must therefore come from some other source. This must be either the back-up material or the joint faces. Examination of the sealant section will usually reveal the origin.

Damaged back-up foam can be a source of blisters. Torn and damaged cells in the foam compress easily under pressure as the joint closes, pushing the air out through the ruptured surface. If the damage is covered by uncured sealant it blows a blister. Repeated movement of the joint acts as a pump, increasing the size of the blister until the sealant becomes sufficiently cured to resist the pressure. Care must be taken to avoid damaging foam backing to minimize this effect.

Porous substrates can also be a source of blisters, especially if the panels forming the joint have been treated with water repellent coating some time prior to sealing. Moisture absorbed into the panels may become trapped behind the repellent surface; this vaporizes when the panels get hot in the sun and the vapour emerges through the edges under pressure, blowing blisters in the seal.

Some uncured polyurethane sealants will react with trapped moisture, generating carbon dioxide gas, which will also cause large blisters in the seal. In many cases the blisters are unsightly but the joint retains a weathertight seal. If resealing is necessary the existing sealant must be removed, the cause of blisters identified and appropriate action taken to overcome the problem changing the sealant if necessary.

2.5.3 Fire resistance

In the event of a change of use of a building, it may be necessary to replace existing joint seals with sealants having an increased measure of resistance to the passage of fire. Such replacement involves complete removal of the old seal and back-up and replacement with new materials. Polyethylene foam may need to be replaced with mineral wool or ceramic fibre materials. A variety of fire-resisting sealants are available including intumescent and fire-resisting ablative materials, and guidance is available from manufacturers as to means of achieving various levels of fire resistance.

2.5.4 Staining

Staining is a phenomenon caused by the migration of constituent materials from a sealant onto or into the surface of adjacent materials, bringing about an observable change in colour or appearance. Staining may be caused by several factors.

Primers

Staining may be caused by primers applied beyond the seal on porous surfaces such as concrete or stone. Most primers for porous surfaces are film-forming and produce a wet varnish-like effect, changing the appearance and giving a dark line effect. Uneven primer application can be unsightly. Some porous surface primers incorporate resins and other products which discolour when exposed to light. Such primers may discolour some time after application, producing brown stains alongside the seal. Removal of primer stains is very difficult or impossible.

Primers cannot be removed from porous surfaces by solvent cleaning. Where primers remain soluble, solvents may spread the stain but in many cases the primer reacts with moisture and becomes insoluble. In many cases the primer film will gradually break down and erode where it is exposed to weathering but this will take several years. Primer stains can be removed from some surfaces by grinding but great care is necessary to avoid damage to the seal.

Reactive constituents in the sealant

Some of the early polysulphide sealants contained phenolic resins which reacted with alkaline materials in cement, producing a purple stain in the concrete. The current British Standard for polysulphide sealants includes a staining test designed to detect this effect. Such tests are not included in all standards and there are no standards for some sealants. Removal of the sealant will not remove chemical staining but in most cases it will prevent it getting worse. It may be possible to remove the stain by chemical treatment but in some cases the stain is unstable and will fade when the source, that is the sealant, has been removed.

Seepage

Most sealants contain oils and/or plasticizers which may migrate into or onto adjacent surfaces. This problem was widely experienced with oil and butyl based sealants but is seldom found with acrylic or polysulphide materials. Problems have occurred recently with both polyurethane and silicone sealants. The most common problem occurs when sealants are applied against slightly porous substrates and the fluid constituents seep into that substrate producing discoloration. Some substrates such as marble or slate are particularly vulnerable to this effect and appear to draw the oils and plasticizers deep into the stone. In general the use of a primer/sealer on the substrate will prevent this effect but it is advisable to test the combination of stone sealant and primer before carrying out the work. Removal of sealant and resealing a joint will not remove the stain. Thorough solvent cleaning of the edge of the stone after removing the sealant may reduce the intensity of the stain or may spread the stain further, diminishing the effect.

In most cases the seepage stain continues to spread for some years. If the source is removed, that is the seal is replaced, the existing stain will continue to spread gradually, reducing in intensity and becoming less apparent. Sometimes the effect may be accelerated by applying heat to the affected area. A less common problem is seepage on non-porous substrates. This is rare with most sealants but has been encountered particularly with silicone products. In this situation fluid from the seal appears to migrate across the surface adjacent to the joint where it collects fine dust on the surface, changing the appearance and creating a stained effect. Contamination can be extensive. This phenomenon is still relatively rare and it has not been investigated adequately to establish the cause, but it is generally associated with highly plasticized sealants and primarily silicone sealants. Removal of this type of staining is extremely difficult. Solvent cleaning reduces the amount of contamination but will not totally remove it. Silicone digester solutions can be used but great care and repeated applications are necessary to ensure adequate removal. After digestion the residue can be washed off with water. The cleaning must be carried out after the offending joint sealant has been removed and special care may be necessary to prevent water penetration through the joints.

3 Examination and inspection

3.1 Identification of causes of failure

Most joint failures occur due to movement and ageing but the underlying cause of the failure may be due to one or more of a variety of defects.

Sealants that exhibit a substantial degree of shrinkage may fail without movement due to the stresses induced by that shrinkage. Joints subject to pressure may fail due to depression of the seal and the stresses imposed on adhesion due to that distortion.

The theoretical amount of movement likely to occur at a joint can be calculated or estimated from a knowledge of the structure and the materials used therein. However, due to friction and other causes, most joints suffer less than the calculated degree of movement. This creates an inherent safety factor. As a result many joints perform satisfactorily for years despite defects in the joint sealing, but where the joint is required to perform to its limits even minor defects may induce early failure.

3.1.1 Inspection

Where joint failure has occurred, it is advisable to examine the project in stages in order to identify the failure and its cause before making a decision on remedial action. An initial overall examination may indicate a pattern of failure suggesting that certain specific details require more detailed investigation. Such a pattern may indicate a structural effect where certain joints in cladding are affected by additional movement, or may indicate problems on certain types of panels or surfaces. If a pattern effect is identified, it is usually possible to carry out a detailed study on a typical area with confidence that the results can be applied to the whole project.

Having completed the overall inspection, attention should be focused onto progressively smaller areas. This more detailed inspection may identify the particular joints or details where failure has occurred. Any pattern effect should be identified and given detailed consideration, especially with regard to movement and/or adhesion to surfaces.

The history of the joint and its surroundings should be checked, together with any surface treatment that may have been applied. Any cleaning operations or treatments that may have been applied should also be checked. Acids and alkalies can damage sealants or substrates, and solvents can destroy adhesion. Bleach or oxidizing agents can damage sealants, especially in their early stages of cure.

The third phase of the inspection is the detailed examination of the joint and the sealant therein.

3.1.2 Visual examination

Visual observation can assist in determining the nature and cause of failure. The position and nature of the failure is important, i.e.

- Adhesion failure – separation of sealant from substrate;
- Surface-cohesive failure – separation just within the sealant, adjacent to the substrate;
- Cohesive failure – separation within the body of the sealant.

The nature of the failure, i.e. torn, waisted or cracked, and the condition of the sealant – crazed, fissured, wrinkled, creased – should be noted.

Where adhesion failure has occurred, separation of sealant from primer or primer from substrate should be recorded as well as the condition of the substrate.

3.1.3 Sampling

Samples should be taken from the joint using a thin sharp knife or scalpel. Care should be taken to cut through the sealant as close to the sides as possible and to cut through sealant and back-up at the ends of the sample so that the sample and backing can be removed from the joint. Wetting the knife blade with detergent solution will reduce sticking. Typical samples should be taken showing the various defects.

Samples should be examined, and the condition of the surface and body of the sealant noted together with consistency, colour, state of cure and any odours released. The dimensions should be noted before wrapping in non-stick material and packing in a box so that they are not deformed.

After samples are removed, the joint should be repaired with a suitable material to maintain the integrity of the building.

3.1.4 Dimensions

Measure the width and depth of the joint seal and dimensions of the joint. Allow the back-up foam to recover and record the size after recovery. Note the prevailing conditions and temperatures at the time of examination.

Note the condition of the substrates either side of the joint and record the presence or otherwise of contamination in the case of adhesion loss.

Note the size, colour and materials of adjacent cladding panels.

Where sealant has separated from the substrate, examine the sealant surface for contamination or debris embedded in the sealant.

3.1.5 Movement

The type, degree and frequency of joint movement must be assessed, either by calculation or by measurement (Section 1). It should however be appreciated that direct measurement can be difficult, time-consuming and expensive.

3.1.6 Sealant identification

Familiarity will enable the observer to distinguish the various types of sealant with a reasonable degree of accuracy, but chemical identification is necessary where identity is critical. Physical characteristics such as elastic recovery, feel, smell and burning odours are frequently adequate to identify the generic type.

Physical characteristics

Most silicones and polyurethanes tend to be fully elastic and will snap back when stretched and released, returning to their original size and shape in a few seconds.

Polysulphides are less elastic, having a slightly deader feel and recovering more slowly when stretched and released.

Acrylic, butyl and oil-based sealants are essentially plastic, retaining their new shape when deformed.

Solvent-based acrylics retain traces of xylene solvent for several years and traces of acrylic monomer can be detected for even longer. Acrylic monomer has an unpleasant pungent smell.

Oil-based mastics have the typical odours of vegetable drying oils such as linseed or soya. These odours tend to diminish as the mastic ages but can usually be reactivated by heating the sample. Mastics become progressively harder with age, losing any ability to recover following deformation.

Burning characteristics

Burning a small sample of the sealant with a cigarette lighter or gas flame produces characteristic results with most sealants. (Warning: emitted fumes may be hazardous.)

Silicones are difficult to ignite. They burn slowly with a fairly clean flame, leaving a substantial white ash.

Polyurethanes burn vigorously with a yellow flame accompanied by dense black smoke, noxious gases and sometimes sooty deposits.

Polysulphides burn slowly, frequently with a yellow smoky flame; however the main characteristic is the choking smell of sulphur dioxide.

Acrylic sealants burn with a smoky flame producing the characteristic pungent odour of the acrylic monomer.

Oil-based mastics tend to burn with a smoky flame and an odour of burning paint.

Butyl sealants usually contain solvents, mineral oils, and sometimes vegetable oils, giving a mixed odour. When burnt, the butyl constituent produces a strong smell of burning rubber which may be modified by the paint odour of the burning oils.

Where products have been in special environments, odours may become modified, particularly where chemical attack or biodegradation have occurred. Foreign odours may also have been absorbed.

Chemical composition

Chemical identification of generic types of sealant is relatively simple in an experienced laboratory. Identification of individual products is much more difficult and in some cases impossible.

3.2 Evaluation

Examination of the sealant failure, bearing in mind the type of sealant, will give an indication of the typical cause of failure.

3.2.1 Adhesion failures

1. Clean separation from substrate.
   Cause: Sealant incompatible with surface (Figure 3.1b).
2. Separation with dirt and debris embedded in sealant surface.
   Cause: (a) Failure to clean surface adequately (Figure 3.1c).
      (b) Surface conditioner or primer needed (Fig 3.1d).

30 *Resealing of Buildings*

(a) Initial state — sealant, polyethylene foam

(b) Clean separation from substrate
    (sealant incompatible with surface)

(c) Separation with dirt and debris embedded in sealant surface
    (failure to clean surface adequately)

(d) Separation with dirt and debris embedded in sealant surface
    (surface conditioner or primer needed)

(e) Separation with surface particles embedded in sealant surface
    (lack of suitable primer to bind or seal surface)

(f) Separation with surface particles embedded in sealant surface
    (very porous surfaces need repeat priming!)

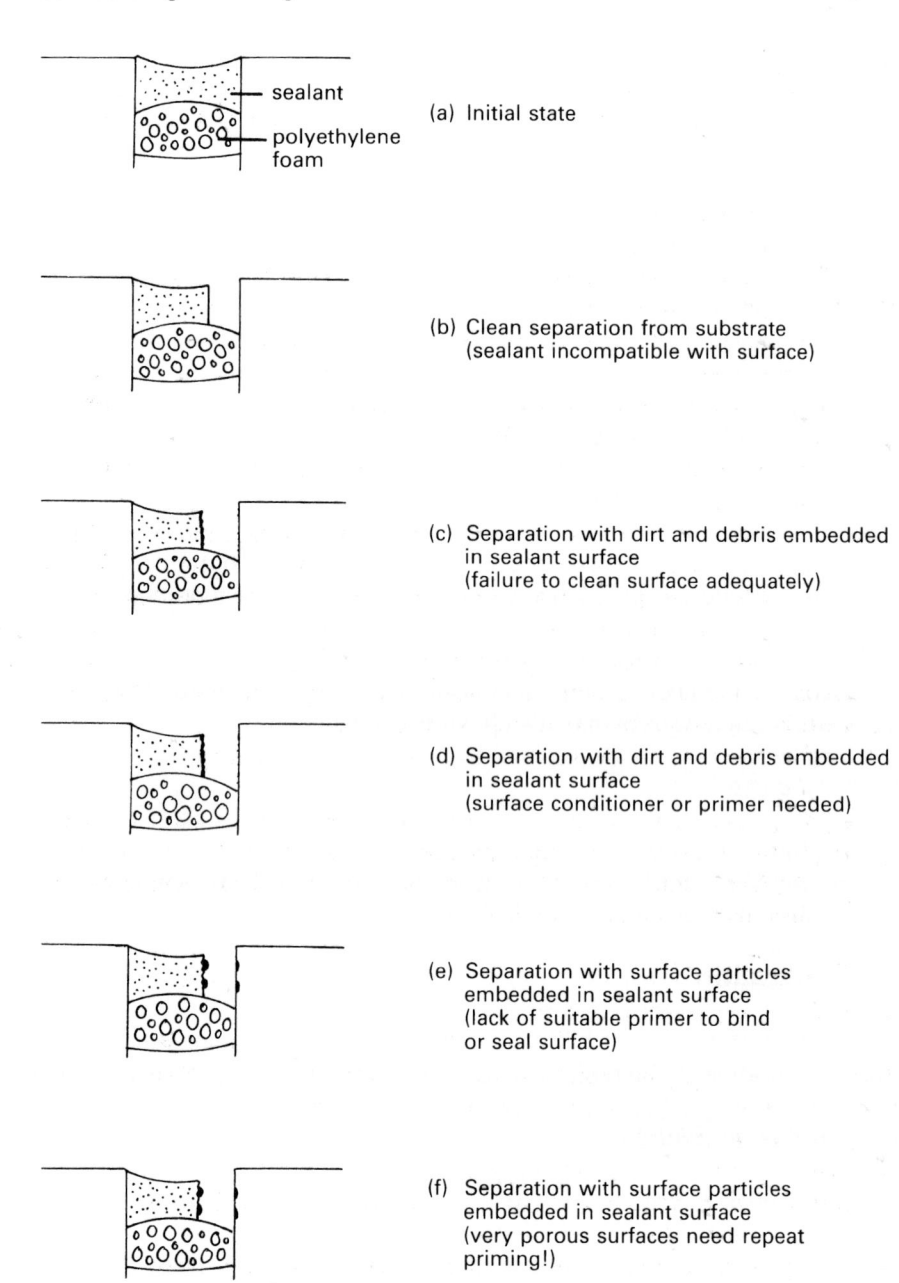

Figure 3.1 *Adhesion failures*

3. Separation with surface particles embedded in sealant surface.
    Cause: (a) Lack of suitable primer to bind/seal surface (Figure 3.1e).
        (b) Very porous surfaces need repeat priming (Figure 3.1f).

3.2.2 Cohesive failures

4. Elastic sealant, central area torn through section.
    Cause: Excessive movement (Figure 3.2a).
5. Elasto-plastic sealant, centre waisted and torn through section.
    Cause: Excessive movement (Figure 3.2b).
6. Plastic or elasto-plastic sealant, folded, waisted and split.
    Cause: (a) Excessive movement (Figure 3.2c).
        (b) Sealant geometry, thin section (Figure 3.2d).
7. Sealant split in middle, mainly plastic or elasto-plastic seals.
    Cause: (a) Back-up folded or laminated, separating as joint opens (Figure 3.2e).
        (b) Rectangular back-up twisted in joint causing thin line of sealant over edge which splits (Figure 3.2f).
8. Two part sealant, soft, sticky centre.
    Cause: Inadequate mixing, unmixed material present (Figure 3.2g).
9. Curing sealant, tacky, paste-like surface, splits.
    Cause: Bio-degradation or chemical attack (Figure 3.2h).

Note Biodegradation occurs in special environments, such as sewage plants, reservoirs and underground conditions.
Chemical attack may result from the use of acid cleaners, bleaches, biocides and similar chemical solutions.

3.2.3 Surface-cohesive failures

Such failures imply splitting or tearing within the sealant close to the substrate, leaving a thin layer of sealant on the surface.

10. Stepped joint, sealant split on plane side.
    Cause: Seal bonded to back of joint, no bond breaker (Figure 3.3a).
11. Sealant split, back-up bonded to sealant.
    Cause: (a) Back-up bonded to sealant, inhibiting stretch of seal (Fig 3.3b).
        (b) Back-up not sufficiently compressed, does not expand to follow movement (Figure 3.3c).

32 *Resealing of Buildings*

Initial state   Failed state

(a) Elastic sealant, central area torn through sealant (excessive movement)

(b) Elasto-plastic sealant, centre waisted and torn through section (excessive movement)

(c) Plastic or elasto-plastic sealant, folded, waisted and split (excessive movement)

(d) Plastic or elasto-plastic sealant, folded, waisted and split (sealant, section too thin)

(e) Sealant split in middle; mainly elastic or elasto-plastic seals (back-up folded or laminated)

(f) Sealant split in middle; mainly elastic or elasto-plastic seals (twisting of rectangular back-up)

(g) Two-part sealant; soft, sticky centre (inadequate mixing)

(h) Curing sealant; tacky, paste-like surface splits (Bio-degradation or chemical attack)

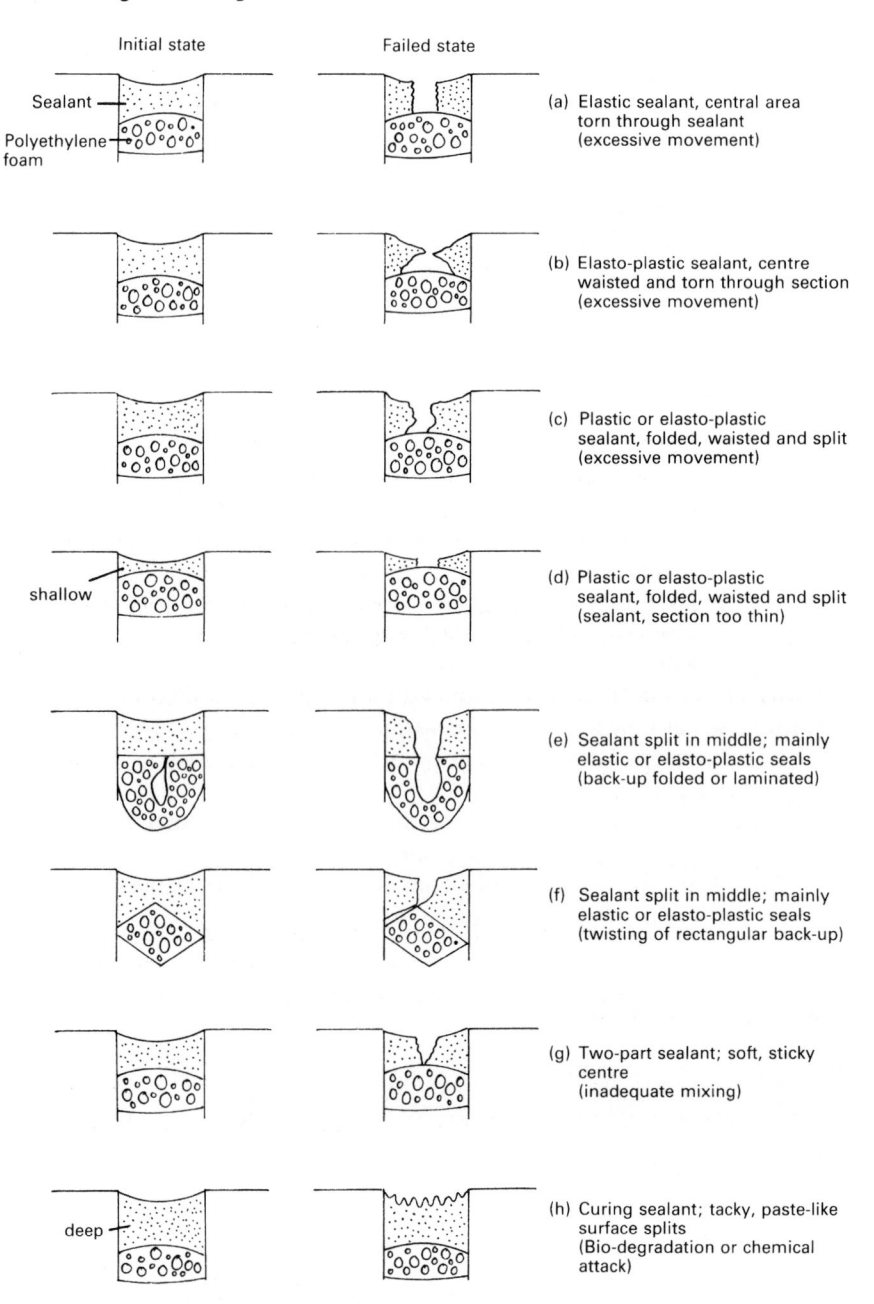

Figure 3.2 *Cohesive failures*

12. Seal torn through at face but separated from substrate behind.
    Cause: (a) Sealant not compacted into joint (Fig 3.3d).
    (b) Substrate not cleaned or primed in depth (Fig 3.3e).
13. Patchy adhesion, some separation. (Figure 3.3f)
    Cause: (a) Inadequate cleaning.
    (b) Surface conditioner needed.
14. Seal torn through on one side.
    Cause: Seal thin on split side, thick elsewhere causing overstress (Figure 3.3g).

3.2.4 Ageing

All sealants tend to become progressively harder and tougher as they age. This results in increased stress as the seal deforms to accommodate movement, placing greater strain on the adhesive and cohesive properties of the sealant. As a result most sealants eventually fail, either by loss of adhesion or by cohesive failure close to the sides of the joint where stresses are high. Some sealants suffer surface degradation due to sunlight causing surface embrittlement, erosion or crazing which can induce cohesive tears and splits.

There are many different formulations of sealants, which may have very different compositions. It must therefore be recognized that the following comments are of a general nature, and that the time and scale of observed defects may vary considerably.

Oil-based mastics

Oil-based mastics initially form a skin around a soft pliable mass of mastic. Ageing causes the skin to thicken and toughen, reducing the volume of pliable material within. The sealant becomes progressively less able to deform. Movement during the ageing process causes the skin to wrinkle and tear, and the mass to deform and waist until failure eventually occurs by adhesion loss or splitting through the hardened mass.

Life expectancy of oil-based mastic is related to volume and depth but can vary between one and ten years.

Butyl sealants

Butyl sealants tend to harden with age but their more critical property is their susceptibility to sunlight. This can cause the emission of a sticky fluid which

34 *Resealing of Buildings*

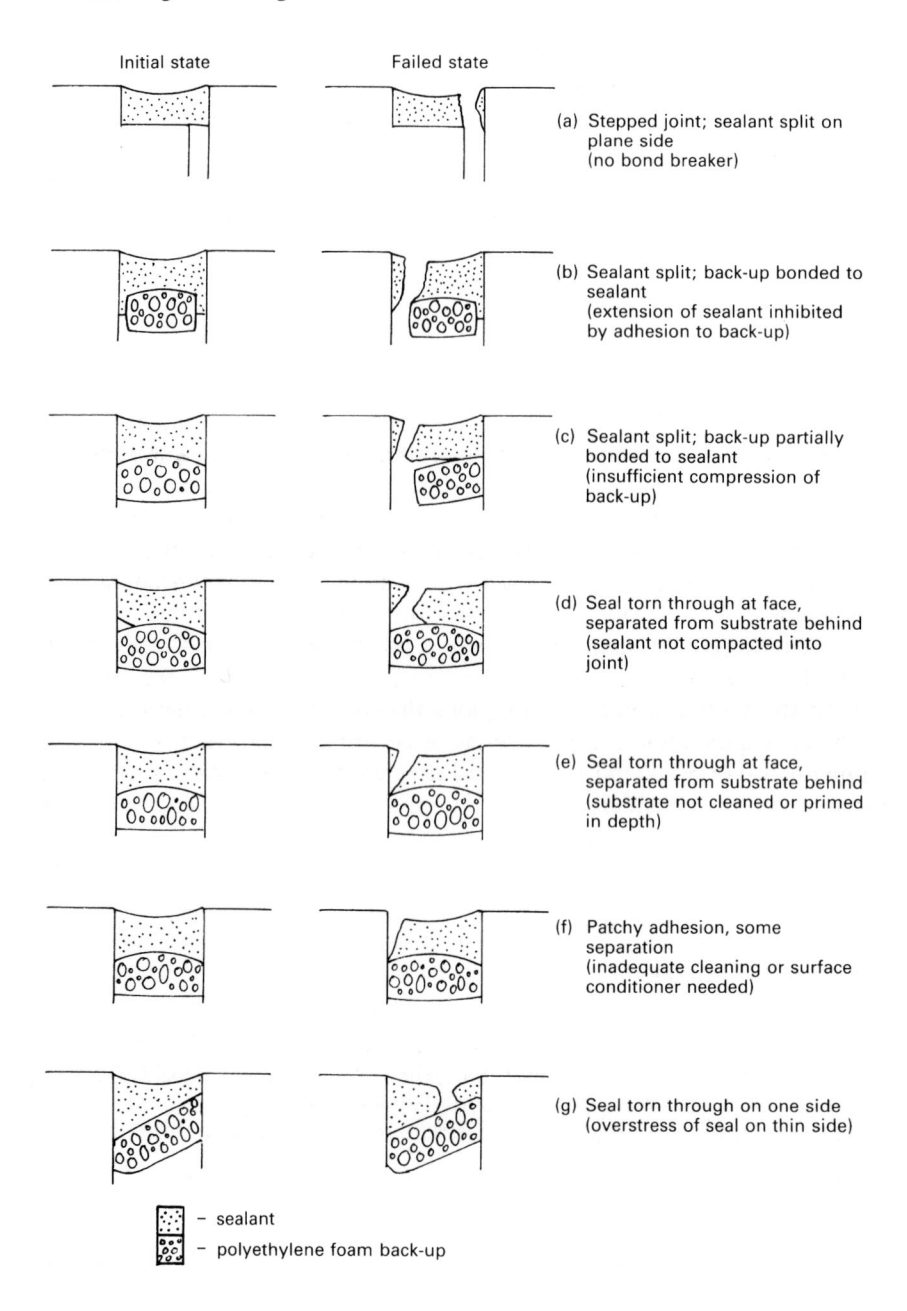

Initial state	Failed state	
		(a) Stepped joint; sealant split on plane side (no bond breaker)
		(b) Sealant split; back-up bonded to sealant (extension of sealant inhibited by adhesion to back-up)
		(c) Sealant split; back-up partially bonded to sealant (insufficient compression of back-up)
		(d) Seal torn through at face, separated from substrate behind (sealant not compacted into joint)
		(e) Seal torn through at face, separated from substrate behind (substrate not cleaned or primed in depth)
		(f) Patchy adhesion, some separation (inadequate cleaning or surface conditioner needed)
		(g) Seal torn through on one side (overstress of seal on thin side)

– sealant
– polyethylene foam back-up

Figure 3.3 *Surface cohesive failures*

collects dust and dirt resulting in a dark furry deposit on the sealant. The result is unsightly and the hardened seal may split .

In some cases sunlight induces the formation of a surface skin which subsequently cracks, causing cracks and splits in the seal.

Life expectancy of butyl sealants is up to twenty years where they are not exposed to sunlight, or one to five years in exposed situations.

Acrylic sealants

Acrylic sealants of the solvent-based type are initially soft and sticky. The surface tack disappears over a few days and the sealant gradually toughens by solvent loss. The sealants are thermoplastic, becoming stiffer and tougher in cold weather. Hardening gradually occurs due to solvent and plasticizer loss.

Some acrylic sealants are susceptible to alkali attack but the more common causes of failure are plastic deformation due to excessive or unsuitable forms of movement, or adhesion failure on friable surfaces.

Water-based acrylic sealants are mainly used internally but can be applied externally, provided they can dry sufficiently before being subject to rain. The adhesion characteristics are generally inferior to the solvent-based products and the most common cause of failure is adhesion loss.

Life expectancy of acrylics is up to twenty years but they are susceptible to frequent cyclic movement such as occurs in curtain walling and light cladding, resulting in a much shorter life.

Polysulphide sealants

Polysulphide sealants cure to an elasto-plastic rubber within a few days, gradually becoming more elastic and tougher as cure is completed and ageing proceeds. Many of the early products toughened rapidly on ageing but modern products have improved considerably in this respect.

Some polysulphide sealants are susceptible to surface crazing. This is a surface degradation and erosion effect. The cracks remain shallow and do not develop into tears or splits. The most common effect of ageing is toughening of the sealant which may induce adhesion failure, particularly on weak or friable surfaces.

Life expectancy of polysulphide sealants is in excess of twenty years.

Polyurethane sealants

Polyurethane sealants vary considerably in their properties as they are based on a wide variety of polymers reacted with various isocyanates. The adhesion characteristics and the ageing qualities vary considerably.

Polyurethane sealants tend to age-harden and some may be susceptible to sunlight degradation and crazing. These effects may induce adhesion failure or tearing.

Life expectancy of modern polyurethane sealants is quoted as being in excess of twenty years. Some of the early sealants failed within a few years but modern materials are considerably better.

Silicone sealants

Silicone sealants are primarily elastic materials although some of the low modulus materials have some stress-relaxation properties, giving them a degree of elasto-plastic character. All silicone sealants are believed to age-harden but the rate of hardening is very slow and is unlikely to be a major factor in failure.

The majority of silicone sealants are truly elastic and therefore joints subject to permanent or long term deformation will induce long term stresses in the seal, increasing the risk of tearing or adhesion loss. Most silicones have a low tear resistance and are susceptible to such effects.

Life expectancy of silicone sealants is generally quoted as thirty years and some of the current sealants have been in service for more than twenty years.

4 Assessment of reseal criteria

4.1 Economics

Resealing a building is usually a cost-driven exercise. Unfortunately it is frequently assessed simply on the cost of the resealing operation itself whereas it should be assessed on cost-in service. *Minor savings in the financial outlay on resealing may result in a substantial reduction in service life, frequent resealing and consequential high maintenance costs.*

The cost of resealing a building is made up of three major parts–access, cleaning and the resealing system. The price of sealant is only a minor part of the overall cost, and hence financial economies in sealant material have a very small influence on the overall cost but could have a major effect on performance.

4.1.1 Perspective on the cost of resealing

The cost of sealing or resealing buildings is made up of various parts and, as a result, *the difference in the total price of a job using a high quality sealant compared to using a cheap sealant bears little or no relation to the difference in cost of the sealants themselves.* The performance of the sealed joint and the durability of the seal are both affected by the sealant, and sealant choice can have a dramatic effect on cost-in-service.

A large proportion of the applied cost is represented by labour, and the largest part of that labour cost is usually joint preparation. *Low price frequently means skimped preparation with consequential risks of reduced joint performance and premature failure.* Economies in joint preparation may result in early failure, giving rise to an enormous increase in the effects on cost-in-service.

The cost of access to carry out the work is usually made up of two parts, that of erecting or moving the equipment and the rental of the equipment. The extra time taken in proper preparation has a very modest effect on the cost of access facilities. Further, the choice of sealant has very little influence on the cost of access to carry out the work.

Although there is no such thing as a typical resealing project, the cost of resealing is made up of:

Cost of access   – usually 5-20% of the total cost
Cost of materials – usually 20-40% of the total
Cost of labour   – usually 40-75% of the total.

Using a better sealant may increase the overall cost of resealing by less than 20% but may increase the life of the joint seal by 100%. The small additional cost is often a very sound investment, showing a substantial return in reduced cost-in-service.

Repair or replacement of sealant in a building generally occurs either because there has been a failure, possibly resulting in water penetration, or because there are other repairs or refurbishment works being carried out which involve resealing. In either case there are frequently economic or time factors affecting the resealing that may affect the quality of the installation. There are many instances where economies in the specification or application process have resulted in substantial remedial costs within a short time.

All too often the time requirements, access needs and weather conditions for proper sealant application are ignored by other contractors. The applicator is called in at short notice and required to complete the work in inadequate time, frequently in unsuitable weather, and often with difficult access or access being dismantled as he works. This inevitably results in poor preparation and poor workmanship as the applicator tries to complete the work to satisfy the contractor.

Sealant application in modern buildings is no longer a case of filling up holes with 'some sticky mastic'. The sealed joint is a critical part of the installation and needs to be treated as a critical component of the cladding, with proper provision in the programme to allow the work to be completed in an appropriate manner to ensure the satisfactory performance of the joints and the entire building envelope.

The following example illustrates a typical case costed at 1992 prices. The building concerned was a small industrial unit, 3 storeys high, with aluminium and glass cladding on all sides. The cladding incorporated about 450 metres of 15 mm wide joints which began to break down after 4–5 years. The building owners were under pressure from their tenants to rectify the problem. The original sealant appeared to have failed due to lack of primer on the joint faces.

Tenders were invited from a number of sealant applicators and offers were received ranging from £2.80 per metre for sealant A, to £3.25 per metre for for sealant B. Both sealants required similar joint preparation and priming but the quotation for sealant B specified a higher degree of surface preparation. Sealant B would have cost about 25p per metre more for materials than sealant A. The building owner accepted the lower tender and the resealing was completed.

Within 3 years the tenant was again complaining of water penetration. This time the joints had failed both adhesively and cohesively, and the building had to be resealed again.

By accepting the lower price, the owner saved £200. The poor performance resulted in a cost-in-service of £420 per year. Properly resealed with a more suitable sealant, having a low modulus and high elasticity, the repair would have cost £200 more but should have lasted at least 20 years resulting in a cost-in-service of only £73 per year. The longer service life should also mean less disturbance for the tenant and less consequential damage and therefore greater long-term savings. The selection of the lower quotation in this case had a serious effect on the cost-in-service and the inconvenience to the tenant.

It is very important when comparing quotations to ensure that the figures are for similar standards of work and materials, and to assess differences and their likely effect on the performance of the joint.

In the above case the initial failure was attributed to lack of priming but, as shown by the subsequent failure, the sealant was not adequate for that size of joint. The greater movement accommodation factor of the silicone sealant would have given an additional safety factor at modest extra cost.

In many cases the difference in price of jobs costed using similar sealants will reflect a difference in standards of preparation, with consequential effect on adhesion. Where different sealants are proposed, sealant cost can be assessed from data sheets and price lists. Substantial differences in the labour element of the quotation usually indicate a difference in preparation standards and attention to detail. Schematic performance scenarios for sealed joints over time are depicted in Figure 4.1.

Resealing is expensive but re-resealing is even more expensive. It is therefore vital that resealing work is carried out properly.

4.1.2 Access

The first, and often the major financial, burden is the cost of access to carry out the work.

40 *Resealing of Buildings*

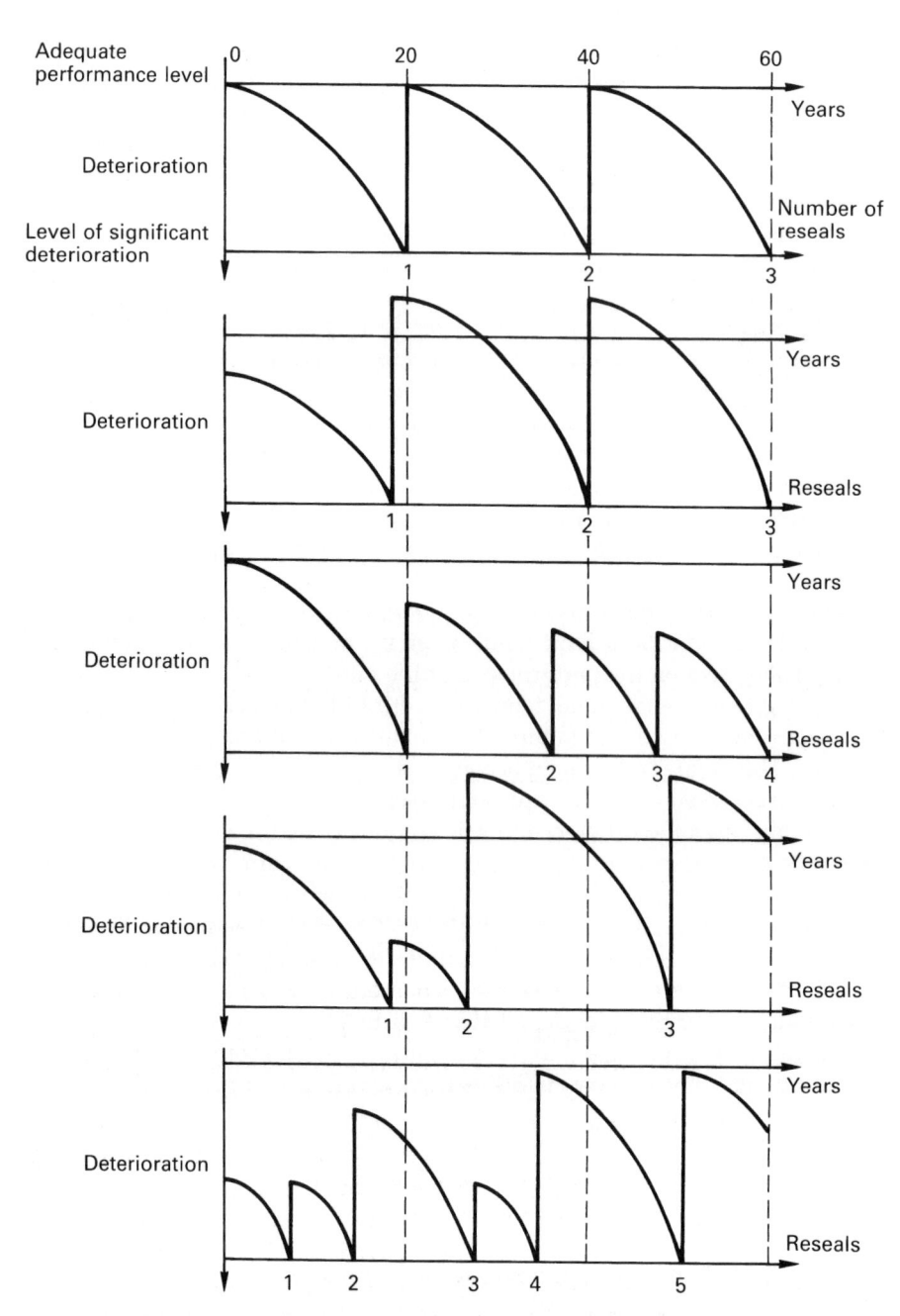

Figure 4.1 *Schematic performance scenarios for sealed joints over time*

Access may involve the provision of scaffolding, cradles, mobile platforms, ladders, cherry pickers or other means of getting to the joints.

Scaffolding is very expensive but usually gives good accessibility and working conditions. If other work, requiring scaffold, is to be carried out at the same time, scaffolding can be used for the resealing operation; scaffolding is seldom used for resealing only because of its high cost. Scaffolding may not always be possible. If, for instance, the area below is not accessible, other means of access may need to be considered. Generally speaking, scaffolding erected for other purposes is often badly placed for joint sealing.

Many tall buildings have their own window cleaning or maintenance cradles and although these may be adequate for minor repairs, they are seldom suitable for substantial resealing operations. In most cases they only give access to a small area of the facade and they cannot carry more than two or three operators or equipment. They are frequently unstable, which leads to difficulties and workmanship faults.

Large cradles or suspended platforms are available, giving access to nine or eighteen metres of facade at a time (Figure 4.2). When properly erected and counter-balanced, these provide a stable mobile working platform to accommodate several operators and their equipment and materials. They also have the advantage that their height is readily adjustable, so that they can be placed at the ideal position for access to joints, whereas scaffold is fixed and not easily varied.

Large cradles are used extensively for resealing work because they are relatively cheap to hire and are usually more convenient for sealant work.

In 1991, when remedial work was planned on a 30-storey tower block in London, the cost of scaffolding was estimated at £200,000. The job was eventually done from large cradles at a cost of £10,000.

In most cases such platforms are suspended from the roof of the building with the counter-balance weights bearing on the roof. They can only be used where the roof is able to accept this load.

Mobile platforms raised on masts alongside the building are an alternative to cradles. They operate in a similar manner to suspended cradles except that they are supported on masts using a ratchet drive system. The masts are self-supported up to a limited height but are attached to the building at intervals above that height. Such platforms generally offer a firm working platform and can also be formed to the shape of the facade, giving good access to complex facades whereas cradles are generally limited to simple rectangular

42 *Resealing of Buildings*

Figure 4.2 *Cradle access*

shapes. The cost of hiring mobile platforms is usually at least twice that of a cradle of similar size.

Another way of obtaining access is the 'cherry picker' or hydraulic platform. This is a small working platform on the end of a hydraulic arm mounted on a mobile unit. The arm, and in some cases the mobile unit, can be controlled from the platform allowing the platform to be raised, lowered, moved sideways or tilted. Cherry pickers are useful for working over low level obstructions and for working against angled surfaces. However, the platforms are usually small giving access to a restricted area, the load-carrying ability is very limited, they are slow in use and very expensive. In 1991, small cherry pickers were available for hire at about £300 per week but larger machines cost £200 per day or more.

Abseil is a new method of access that has recently been proposed. This technique has been used for inspection purposes but does not appear suitable for working on the joints. The amount of cleaning and application equipment needed is more than can be carried conveniently, and it is unlikely that scraping, scouring and solvent cleaning could be carried out efficiently from an abseil rope. Priming, caulking and sealing would require repeated visits to the joint, making this method of access less efficient.

The bosun's chair has been used as a means of access in special circumstances. It is more suitable than abseil but is only suited to small sections of jointing because of the limited facilities available.

Both abseil and bosun's chair techniques require special skills and would also require special consideration with regard to Health and Safety at Work, both for those involved and those working or passing below.

Ladders may be adequate for small amounts of joint repair on two- or three-storey buildings, but the low cost is often offset by the limited area of accessibility and the consequential frequent movement and rearrangement necessary. Mobile tower scaffolds are often to be preferred, giving a more stable working platform and a greater area of access.

4.1.3 Resealing options and cleaning techniques

The second major factor affecting the cost of resealing is the cost of removing the old seal and preparing the joint to accept the new seal.

Replacement techniques may involve three options:

1. Complete removal of the old seal and complete replacement with similar or different material. It has been proved within the RESEAL project that this option confers the best performance (Section 5.2.2).
2. Removal of the bulk of the old sealant and replacement with new sealant applied onto the residue of the old. Careful experimental work has proved that the resultant performance is reduced, in proportion to the type and amount of residue. Non-curing contaminants are most unlikely to provide a suitable surface.
3. Oversealing.

Complete replacement

In most cases Option (1) is intended, that is complete removal of the old sealant and complete replacement. This is the preferred choice for optimum performance, especially if the original seal has failed by loss of adhesion. Such failure indicates that the adhesion is suspect even in areas that have not failed, and where failure has occurred the surfaces will have been contaminated by dirt washed into the joint.

Complete removal of the existing sealant may be difficult and tedious. Accessibility of many joints is often less than easy. It may be difficult to get tools into the joint to cut or scrape the sealant from the surfaces. Abrasion by angle grinders may be possible on stone or concrete surfaces if the joint is wide enough. However, this is often not practical if one side of the joint is formed by a material or surface finish vulnerable to mechanical damage such as an aluminium window frame.

Rotary wire brushes are also limited in their applicability because they can easily damage many materials. There is also a risk that such tools will spread contamination, particularly with soft or thermoplastic materials. Scraping and brushing are the most common methods of abrasive removal of sealant.

Abrasion may not be suitable in some cases because many older sealants may contain asbestos fibres which may be released into the air by grinding or burning.

Chemical or solvent cleaning is frequently restricted by the materials forming the joint and/or materials adjacent to the joint. Most of the solvents that attack sealants also affect paint finishes and polyesters. Many concrete or stone claddings are coated on the reverse side with bituminous materials. Bituminous or pitch-based dpc's often penetrate the joint areas. Many solvents attack bitumen or pitch-based materials, spreading the brown stain into porous cladding materials. The use of solvent may spread the contamination across large areas of the joint surface and adjacent surfaces.

With materials such as stone or concrete, the primary means of cleaning is scraping and abrasion followed by careful solvent cleaning.

Anodized aluminium and polyester finishes are easily damaged. The bulk of the sealant can be removed by cutting and scraping with a plastic scraper, and the residue removed with solvent assisted by a nylon or polypropylene scouring pad if necessary. Silicone residues can be removed with a silicone digester.

Despite the best efforts, it is often found that some traces of sealant will remain securely bonded to the surface. Provided those residual traces are small and the new sealant system is compatible, this may not be detrimental.

Partial removal and replacement

Option (2), partial removal and replacement of the seal by applying new sealant onto the residue of the old, is not often practical unless the seal failure was cohesive. If the original seal is firmly bonded to the sides of the joint, it may be possible to cut out the old seal leaving 0.5 mm or less of sealant firmly bonded to the joint face, and then to apply new sealant onto the old. Obviously the new sealant must adhere to the old and that must remain bonded to the joint face. Where new sealant is to be applied to old existing material, tests should be carried out to ensure compatibility and adhesion. It has, however, been proved that a sacrifice in performance will be made without complete removal (Section 5.2.2). (*See* Appendix 2.)

In addition, the new sealant must have a lower modulus than the old otherwise any movement will impose greater stress on the existing sealant causing rapid joint failure.

This technique is generally used where failure was totally cohesive but it can also be used where failure was adhesive and confined to one substrate only, especially where the joint is wide enough to allow cleaning and preparation of that surface without disruption or contamination of the sound adhesion surface.

Oversealing

Option (3), oversealing, is not often practical but where it is possible it may offer a cheap, easy and usually reliable resealing system.

With deeply recessed joints, oversealing may be an easy option but there must be sufficient depth available to allow cleaning, preparation and priming if necessary without picking up the old sealant and contaminating the new bond area. The joint area must be deep enough to accommodate a back-up foam or bond-breaker tape together with an adequate depth of new sealant.

Oversealing can sometimes be achieved by creating a new joint in front of the old. One method is by extending the joint faces, bedding metal angles or timber fillets onto the face of components either side of the joint, effectively creating a new joint area in front of the original. Another method is by applying a cover plate or butt-strap over the existing joint and forming a joint in shear alongside, in front of the original. This technique has been used for perimeter seals to windows, for joints between panels and mullions, as well as panel cladding to plant rooms and for copings. This method has advantages where cleaning the original joint is excessively difficult, or where the existing joint is too small and cannot be enlarged.

Another form of oversealing is the bandage joint. In this case the new seal is formed over the face of the original joint, lapping onto the face of the cladding either side and forming a thick band of sealant bridging over the original seal and separated from it by a bond-breaker tape, but bonded to the panel face either side. This form of repair has been used extensively on copings, plant room cladding and similar situations where the joints are not easily damaged and their appearance is not critical. Typical oversealing details are shown in Figure 4.3.

Oversealing over fillet seals is generally undesirable. It results in very large fillets as the new seal requires a 6–10 mm bond to fresh clean surface either side of the joint, often resulting in fillets 20 mm wide or larger.

46 *Resealing of Buildings*

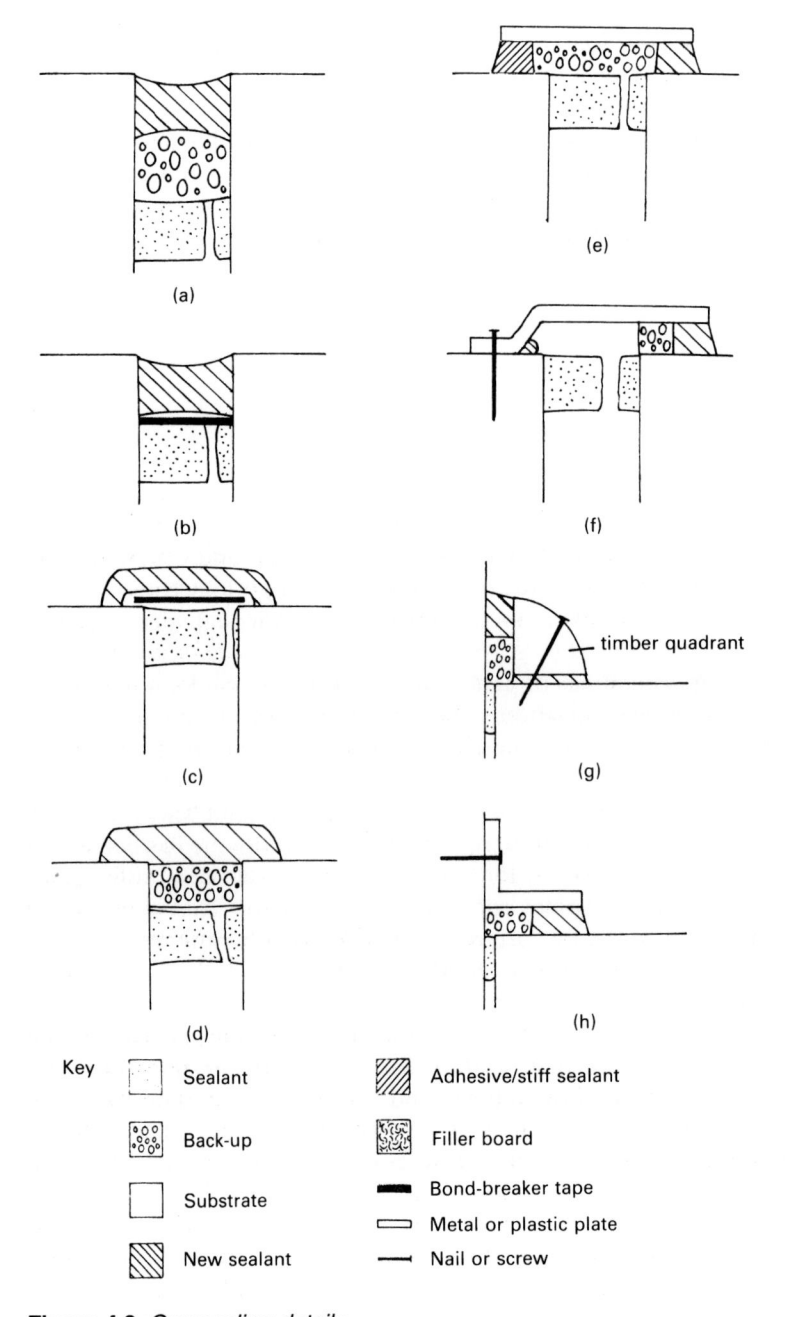

(a)

(b)

(c)

(d)

(e)

(f)

(g) timber quadrant

(h)

Key
- Sealant
- Back-up
- Substrate
- New sealant
- Adhesive/stiff sealant
- Filler board
- Bond-breaker tape
- Metal or plastic plate
- Nail or screw

Figure 4.3 *Oversealing details*

Applying new sealant direct onto failed sealant is undesirable because the failure in the old will be transmitted into the new causing early breakdown. The introduction of a bond-breaker tape or foam rod is recommended to separate the seals but this frequently leads to even larger fillets.

In most cases complete removal of the existing seal is desirable but because of the difficulty of removing all traces it is advisable to reseal with a compatible sealing system to minimize the effects of such traces.

4.1.4 The sealing system

The third portion of cost is the sealing system itself–sealant, primer, back-up foam, etc.

In view of the high cost of access and cleaning, the priority in choosing the sealant system should be ensuring the best possible choice to minimize the need for resealing again. The sealant should be selected to have the longest life commensurate with a high probability of successful application to achieve that performance and to suit the expected life of the building. This should ensure that future resealing will not be necessary for a very long time but it must be appreciated that this may be less than the life of the building.

The cost of sealant and the cost of priming are relatively small compared to the cost of access and cleaning. There is little point in economizing on these if it puts the job at risk, or is likely to result in an early or accelerated failure.

Tests on concrete and stone have shown that primers improve adhesion, and that adhesion is maintained for a longer time especially when the seal is subject to damp or wet conditions. Age-hardening can also place great demands on adhesion and give rise to premature failure if a primer is not present. Omission of the primer could be a false economy and may reduce the life expectancy of the seal by more than 50% or could even result in immediate failure.

The specifier must select the most suitable sealant for the job. The sealant should be chosen to accommodate the expected type, frequency and amplitude of movement. Life expectancy is also an important factor. It may be possible to meet the basic requirements with a number of different sealant materials. Selection may then be considered on the basis of compatibility, joint cleaning requirements, and so on, which will affect the probability of achieving a satisfactory seal. Sealant performance may also be a factor, particularly in terms of safety factors as joint movement calculations are not always accurate and not all conditions predictable. Cost-in-service and the price of the overall job may also influence the final choice.

Compliance with national standards is seldom an adequate specification for the sealant. Sealant manufacturers compound their sealants to their own formulation, based on their own experience and their own particular market section. As a result there are considerable differences between individual products although they may comply with the same standard. For instance a manufacturer who mainly supplies civil engineering contractors may make his sealants tougher than a manufacturer supplying the curtain wall industry. Both sealants comply with the standard but fit opposite ends of the specification to suit their normal uses. Interchanging those sealants can lead to disaster.

Specification testing is usually carried out on clean fresh surfaces which are easily reproduced in laboratories. Standard contaminated surfaces are very difficult to reproduce but the good manufacturer knows where and how his sealants are used. He builds a library of experience and may modify his product to meet service conditions, still meeting the standards but giving his product an individuality that makes it different to others. This individuality is particularly apparent in the ability of a sealant to tolerate contamination. Some sealants will displace or tolerate contamination whereas others require a high degree of surface cleanliness; both sealants may comply with the relevant standards and performance should therefore be established by trials.

Any defects or deficiencies in joint cleaning and preparation are likely to reduce the adhesion strength of the sealant/substrate bond. Using a low modulus sealant reduces the stress on the adhesive bond. Where the choice is available, selecting a lower modulus sealant can give an additional safety factor. However, this option is not usually available for joints subject to pressure or traffic.

Selecting a sealant that is compatible with the previous sealant system is desirable in order to minimize the risk of interaction with any residues that may remain in the joint. Similarly, if primer is used, the primer should be compatible with any seal residues and be equally effective on clean and contaminated areas.

If the reseal system is not compatible with the original system, all traces of the original MUST be removed prior to resealing.

Back-up materials

Back-up materials frequently need to be replaced. Foam materials are usually removed or displaced with the sealant. If they are left in place, they tend to obstruct access for cleaning and priming and they are likely to be contaminated in the process. It is sometimes easier to push the old backing foam to the back of the joint to isolate contaminants and reduce the risk of spreading contamination onto the joint faces. In cavity construction, care should be taken to ensure that the back-up foam does not fall into the cavity or form a bridge.

In many cases the back-up foam will need to be replaced with the correct size in order to ensure optimum seal performance. The foam backing should be tight in the joint to give firm support to the sealant, it should be compressed such that it will expand to follow the movement of the joint during cure and be thick enough to prevent buckling as the joint closes. Care should be taken to ensure that the backing foam is at least 20% wider than the joint and that the thickness is at least half the width.

Back-up foam should be such that the sealant does not adhere when cured so that the seal is bonded to the opposing faces of the joint only, allowing the sealant to stretch or compress without impediment. Polyethylene foam is extensively used but should either have a skin on the surface against the sealant or should be cut from material having a fine closed-cell structure. This ensures that the sealant releases from the surface easily whereas sealant applied to the cut surface of coarse cell foam penetrates into the cells and interlocks with it, inhibiting release and preventing free movement of the sealant.

Other compressible materials may be used as sealant back-up in special situations. A very low density open cell foam has been used to prevent blistering in panel joints (*see* Section 2) but should be used with caution as it absorbs water readily. Expanded rubber materials are used in load bearing joints but may cause staining and discoloration of some sealants.

Where polystyrene or fibreboard are used as joint fillers, a bond-breaker tape should be applied to prevent sealant adhering to the filler board.

The new back-up material should be positioned to give the appropriate depth of sealant to suit the sealant type and the application.

4.2 Selection of the sealant system

The specification for resealing the building will depend on the factors of access, cleaning, sealant and overall cost. It frequently represents a compromise in that the type of sealant may not be ideal and may have to be selected on the basis of its tolerance to conditions prevailing, rather than the ultimate product which may require clinical conditions. A selection scheme is provided in Figure 4.4, whilst the headings for appropriate Survey, Specification and Assessment sheets are provided in Appendix 1.

50 *Resealing of Buildings*

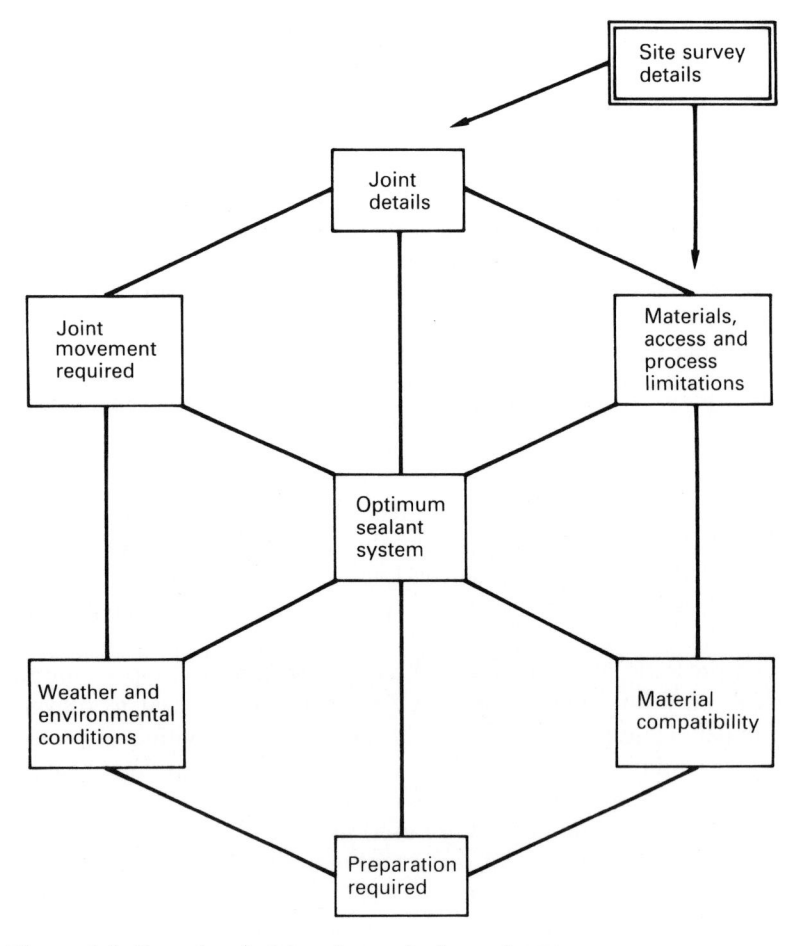

Figure 4.4 *Reseal material system selection scheme*

4.2.1 Movement considerations

In assessing the sealant it is advisable to look at the building as a whole and see if there is any pattern of failure which indicates either special service conditions or some error in the original design. It is not uncommon to find specific areas of failure which indicate some abnormal situations affecting individual joints or groups of joints.

Situations such as major failure on the south or west elevations are frequently associated with the severe sun and rain weather conditions on those elevations, whereas north and east elevations tend to suffer less driving rain and are sheltered from the worst effects of the sun. General failure on a particular elevation may therefore be associated with weathering effects.

Other situations, such as the top of the building being more severely affected than the bottom again may indicate a weathering effect rather than a structural variation. Other pattern effects, particularly linked to groups of joints, may indicate special factors linked with the construction detail.

Pattern Effects

The following two examples illustrate pattern effects associated with groups of joints.

> ***Case study 1: large office block clad with storey-height pre-cast stone units hung on to pre-cast concrete frame.*** *Each stone unit comprised a mullion section, soffit panel, and window surround. All units were identical and sealed vertically and horizontally with a one-part polysulphide sealant.*
>
> *After about two years it was noted that every third vertical joint was showing severe signs of distress. This was worst on the southern and western elevations but noticeable on the north and east. The two joints between showed no signs of distress or severe disturbance. All joints were the same size. Detailed investigation showed that in the building's construction three panels were hung from the same pre-cast concrete beam at each floor level. The beams were fixed at one end and allowed to slide at the other. The third joint at the end of the beam, was therefore subject to thermal expansion of the beam as well as the individual panel. The intermediate joints only suffered the movements of the panels, and this was probably relieved to some extent by the beam's expansion. The beam was protected against direct radiation from the sun by the cladding panels but would probably suffer movement akin to ambient temperature changes, thus substantially increasing the diurnal movement of the third joint. Resealing was therefore confined to the third joint, which was resealed with a low modulus silicone sealant in place of the polysulphide. The greater Movement Accommodation Factor (MAF) was adequate for the increased movement.*

Case study 2: large office block comprising a tower eighteen storeys high and a podium two storeys high, clad in black anodized aluminium sealed with a two-part polysulphide sealant. After one year the sealant on the podium failed whereas the tower remained completely sound. Detailed investigation showed that the tower was clad with a poorly insulated curtain wall system which formed the total external cladding. The podium was clad with a similar curtain wall system but it was erected to cover a heavy in situ concrete structure and the space between packed with fibreglass to prevent panel drumming and vibration.

Cladding on the tower showed modest expansion and contraction because of the poor insulation standards. On hot days heat absorbed by the cladding was dissipated into the building and when the external conditions were cold, heat from the building kept the cladding warm. The cladding therefore showed modest expansion and contraction.

Cladding on the podium was well insulated by the mass concrete structure behind and the fibreglass packing. The cladding could not transmit absorbed heat to the building because of the insulation and neither could it receive heat when the external conditions were cold. The cladding on the podium therefore showed a much greater degree of thermal expansion and contraction and a much faster reaction to changes in external weather conditions. It suffered larger, faster and more frequent cycles of movement. Detailed design analysis showed that the polysulphide sealant was not adequate to accommodate this pattern of movement and the sealant was replaced with a more elastic, lower modulus material which subsequently gave satisfactory service for many years.

Pattern-of-failure effects therefore show joints subject to abnormal movement and detailed evaluation of the construction may identify the cause of that movement and premature failure, and indicate the method of achieving a long term solution to the problem. Where more general failures occur, examination of the individual failure, outlined in Section 3, may give an indication as to the cause and hence some indication as to the remedial treatment required.

If failure is due to ageing, that is the sealant has performed for a similar period to its life expectancy and that period is considered reasonable in relation to the life of the building, it indicates that the original choice of sealant was satisfactory; resealing may then be carried out with either the same material or possibly an alternative material with improved durability.

The choice of replacement sealant would depend upon the comparative costs of replacing with the same sealant and replacing with the alternative.

Replacing with the same sealant may be justified if the replacement sealant is compatible with the old, reducing the requirements for cleaning and preparation. However, old and new sealant may not be compatible. With some sealants, such as some silicones and some polyurethanes, new sealant of the same type will not adhere very well to old. With long-established products, there may have been formulation changes that affect the compatibility of new and old. Where the alternative sealant only involves the same cleaning and preparation it would seem logical to use the more durable material.

Where there is an increased cost of cleaning or preparation, consideration should be given to the cost-in-service, taking into account the total cost of the remedial operation and the resealing frequency requirements.

Where sealant failure has occurred prematurely consideration should be given to the cause of that failure, that is whether it is due to design defects or poor application. If failure is due to an application defect it may be adequate to replace with the same sealant correctly applied. However, the fact that failure has occurred shows that the joint is moving and it may be worth considering an alternative sealant with increased performance in order to reduce the stress on the sealant. If the sealant is being called upon to perform to its limits it indicates a need for 100% perfection in application. Such a quality of workmanship may be difficult to achieve in practice, especially in remedial work. The use of an alternative sealant may be less critical in application and improve the likelihood of a satisfactory installation.

Where seal failure occurs due to a design fault, e.g. excess movement, it is necessary to analyze the situation prior to selecting the type of remedial sealant. Such an analysis should include assessing the movements originally affecting the joint and reassessing the movements currently affecting the joint. Replacing the sealant with a product having a higher Movement Accommodation Factor may not be sufficient, it may be necessary to enlarge the joint, or take other measures to ensure that the performance of the joint is within the capability of the seal.

Many joints in buildings are subject to a variety of movements from different causes. Some of those movements may be permanent deformations, such as settlement, drying out shrinkage, creep, and so on, whereas others, such as thermal expansion, are repetitive cyclic movements. When resealing a building the permanent or long term deformation effects may have already taken place or be substantially complete; in such cases the allowance for such movements can be substantially reduced when assessing the reseal requirements.

Drying-out shrinkage on concrete and many other materials is generally substantially complete within two to three years after completion. Concrete creep takes longer but in most cases it is about 90% complete within fifteen years. Settlement in most buildings is substantially complete within about five years but care should be taken to check the type of subsoil and the effects of changes in water table which may induce movement long after this period. Brickwork expansion is normally substantially complete within five years. These factors can therefore be eliminated or reduced considerably when considering resealing joints in buildings more than five years old. Some reduction may be appropriate on even younger buildings.

In considering movement it should be appreciated that the joints may be permanently changed in size due to the effects of these permanent deformations. Some joints may be substantially larger than the original construction size whereas others may be narrower. It is necessary therefore to determine the size of the joints in order to assess the minimum Movement Accommodation Factor (MAF) required for the reseal material.

In assessing the joints for resealing therefore, the degree of movement and the amplitude of movement may have changed. The overall character of movement may also have changed in that the permanent deformation movements would already have taken place, and most of the remaining movements would be of a cyclic nature and therefore requiring elastic or elasto-plastic sealants rather than plastic, stress-relieving materials.

Reassessment of the joint should therefore indicate the nature of the movement to be accommodated, that is cyclic or permanent deformation, and the relative combination indicating the need for elasto-plastic or elastic sealants. The Movement Accommodation Factor (MAF) will indicate which of the appropriate sealant materials are adequate for the purpose and should produce a short-list of suitable materials.

4.2.2 Cleaning and access

If the joints are easy to clean and all traces of the sealant system can be eliminated, the choice of sealant is open and the most suitable for that application can be selected bearing in mind durability, cost, and so on. If on the other hand cleaning is not easy and there is a probability that some contamination from the old sealant system will remain, compatibility becomes a serious consideration and the eventual sealant choice must be considered in terms of compatibility with the residue of the original seal. *It must also be accepted that a reduction in performance is inevitable without complete removal of the old sealant system* (Section 5.2.2). If a compatible sealant is available which will bond satisfactorily to the residues of the original sealant, which in turn is well bonded to the substrate, sealing onto that residue is a possibility. If, however, suitable sealants are not compatible with the residue of the original sealant system, that residue must be completely removed in order to achieve a satisfactory installation. The ease or otherwise of removing that original sealant system is a critical factor. In considering the removal of the original sealant access may be a further factor. Scaffolding is generally a firm secure platform on which to work making it practical to use grinders, rotary brushes and similar power tools assisting removing the old sealant. Cradles and mobile platforms vary in rigidity and firmness depending on size and means of retention against the face of the building. In many cases they do not offer a firm enough platform to use heavy power tools and therefore where cradles, or similar platforms, are used there will be restrictions on the type of grinding tools that can be utilized.

In most cases the use of heavy or violent cleaning tools is not practical. In a few cases grit-blasting or grinding may be possible, but in the majority of situations cleaning is restricted to cutting out, scraping, and solvent cleaning. The general tools used for cleaning include scrapers, knives, chisels, wire brush, stiff bristle brush, plastic or metal pan scourers, rags, and similar cleaning cloths. A good supply of clean cloths for final wiping is essential. Solvent cleaning may be restricted if any of the surfaces or materials adjacent are susceptible to solvent attack (*see* Section 5). Un-cured sealants are liable to be dissolved and spread by solvents, especially on porous surfaces.

The efficiency of cleaning often relies on the operator and his imagination and determination. Regular inspection of the work at the cleaning stage is desirable to ensure that satisfactory standards are maintained.

4.2.3 Weather and environmental conditions

All sealants are affected by environmental conditions. Temperature, humidity, moisture and other factors affect the behaviour of sealants and can affect their suitability for specific projects.

Temperature

High temperature will tend to reduce the viscosity of most sealants and low temperatures will increase the viscosity, making the sealant stiffer and more difficult to mix or to gun. Different polymeric materials are affected to different degrees and individual formulations of similar generic types may differ. In general the viscosities of silicone sealants are least affected by

differ. In general the viscosities of silicone sealants are least affected by changes in temperature whereas acrylic and bituminous materials are affected to a considerable extent.

Most sealants are poor conductors of heat and, once cooled, they take a considerable time to warm up. It is advisable to store sealants at a convenient temperature to avoid unnecessary difficulties in application.

Two-part sealants are affected in two ways. At low temperature, the increased viscosity makes the sealants more difficult to mix and imposes increased strains on mixing equipment. At high temperatures the cure rate is increased, reducing the work life of the sealant, which may make it difficult to use the mixed sealant before it becomes too viscous to apply.

The cure mechanism of some two-part sealants is exothermic, (that is, they generate heat) and thus accelerates the rate of cure. Such sealants should be mixed in small quantities that can easily be used within the work life.

The cure of one-part sealants is initiated by solvent loss or atmospheric moisture and is less susceptible to changes in temperature, but is generally much slower than two part sealants.

Moisture

Moisture vapour in the atmosphere initiates the cure of most one-part sealants. In the UK relative humidity is generally above 50% and is ample for all types. Silicone sealants will cure at less than 25% RH whereas polysulphide sealants generally require about 30% RH to achieve a satisfactory cure rate.

There are a few sealants that can be applied to wet surfaces or under water but these materials are very specific in their application. In most cases sealants should be applied to clean dry surfaces as the presence of water will prevent good adhesion. Most sealants should not be applied in rain as the presence of droplets of water on the surface could inhibit adhesion. Some sealants have a greater tolerance of moisture than others but such tolerance is usually very specific to particular surfaces and does not apply in all cases.

Frost-low temperatures

The major hazard of sealing at low temperatures is the danger of frost or ice on the surfaces. When temperatures have been low overnight surfaces can remain cold long after the air temperature has risen, resulting in a thin layer of ice or condensation remaining on surfaces that could prevent the sealant sticking to the surface. Special care is necessary to remove this moisture and prevent it from reforming.

Great care is needed at temperatures between +5°C and −10°C when condensation and frost can form. At very low temperatures there is very little moisture in the atmosphere and hence less danger of condensation.

Special environmental conditions

Where special environmental conditions apply, such as swimming pools, saunas, hospitals, breweries, cheese factories, abattoirs, factories, etc, special sealants may be necessary. Particular properties may be needed such as cleaning resistance, mould resistance, health and safety factors, compatibility with processes, and so on, which must be taken into consideration when choosing the reseal system.

4.2.4 Check list

In choosing the sealant system the following check procedure should be followed by the specifier or contractor:

1. Identify the type and causes of movement affecting the joint.
2. Check the type of movement and select the sealant type appropriate, i.e. elastic, elasto-plastic, etc.
3. Check the joint size and amplitude of movement to assess the minimum Movement Accommodation Factor (MAF) required.
4. Check surface conditions required for each suitable sealant.
5. Check whether surface conditions can be achieved for each sealant system, i.e. cleaning required priming, compatibility, etc.
6. Short-list possible sealants and specification; check cost and life expectancy.

4.3 Site trials

After selecting the preferred sealant system or systems it is advisable to carry out site trials before undertaking the complete resealing of the building. Such site trials may reveal unexpected difficulties or other factors that may affect the resealing decision. Many buildings will have undergone surface treatments during their life – paints, water repellents, and so on having been applied. These may affect adhesion and other factors and may not be apparent during initial examination. Site trials should indicate whether it may be necessary to vary the sealant system to accommodate the residual effects of such treatments.

An example of such problems may occur with buildings treated with water repellent materials. Where these are applied over porous surfaces, they are liable to lead to a build-up of salts and other agents in the surface just below the water repellent treatment. The water repellency deteriorates with age and water soaks into those surfaces relatively quickly and easily. It does not escape through the surfaces as quickly because the insoluble salts below the surface impede the loss. This can lead to the build-up of a high moisture content within the panel which takes some time to dissipate.

The cleaning and preparation of the edges of the panel may disrupt the water repellent/salt layer, or the edges may have been untreated, resulting in a relatively easy moisture egress from this area. Such moisture can interfere with the adhesion of sealants and can also react with such materials as polyurethanes, causing blistering of the seal. Correct surface preparation and priming can avoid this situation.

Panels subject to high temperatures may reveal the problem in another way in that the moisture may be expelled as moisture vapour which can again cause blistering of the sealant unless the jointing area is sealed with two or three coats of primer/sealer prior to sealing the joint.

Invisible traces of silicone or some other contaminants in the jointing area could interfere with the adhesion of many sealants. Adhesion should be checked when the sealant has cured to a substantial degree.

Unless the history of the building is known in detail and confirms that there is no risk of any detrimental action, it is always advisable to carry out trial sealing with two or three different sealant systems prior to undertaking the resealing of the building.

4.3.1 Adhesion tests

When sealants cure, they either form a skin or firm through to form a rubbery mass. Adhesion can be examined qualitatively by peeling the sealant away from the substrate by finger pressure. The adhesion should be greater than the cohesion as is indicated when the sealant tears through the mass rather than peeling from the substrate.

In most cases it is possible to remove samples of the joint substrates and submit these to the laboratory where tests can be carried out including water immersion tests. Where it is not possible to remove samples, adhesion tests can be carried out on site, including water immersion in many cases.

The test specimen should be prepared and primed, if necessary. Sealant should be applied either as a bead 3–5 mm thick on the face of the test piece, or as a sealed joint between panels and allowed to cure under normal site conditions.

Adhesion can be tested by pressing the sealant near the interface to peel the sealant from the substrate. Pressure should be increased gradually until the sealant begins to separate from the substrate. If a satisfactory seal is to be achieved, the bond strength should be greater than the cohesive strength of the sealant and the sealant should tear before it peels from the substrate.

If a satisfactory bond is obtained under dry conditions, the sample should be soaked with water and the adhesion reassessed.

Soaking in water can be achieved either by spraying the specimen continuously or by forming a container incorporating the specimen and filling the container with water (see Figure 4.5). The wet conditions should

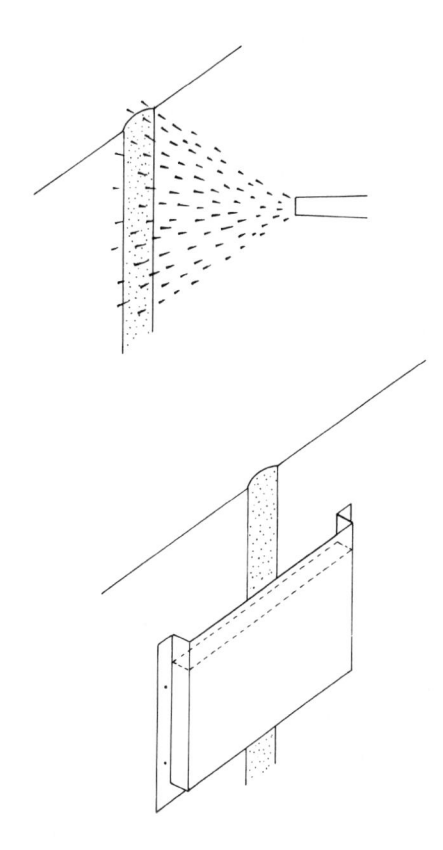

Figure 4.5 *On-site water-wetting tests*

be maintained for at least 24 hours and preferably 7 days. On completion the sample should be reassessed immediately after the water is removed, whilst still wet. The bond is considered satisfactory if the adhesion is greater than the cohesion so that the sealant tears and does not separate from the substrate. However, the sealant should be adequately cured before carrying out this test.

5 The reseal operation

5.1 Access

In most cases access will be from ladders, suspended cradles or working platforms. In a few cases access may be from scaffolding, but as scaffolding is normally erected for other purposes it may not be placed conveniently for the sealant applicator. Scaffold platforms frequently coincide with joint positions making access difficult if not impossible. Dismantling and reassembling the scaffold is expensive, and in some cases it may be more convenient to remove the scaffold and substitute large cradles to carry out the resealing operations.

Difficult working conditions may require special means of access and may also influence the choice of sealant. Inclined surfaces, roof areas and glazing may create special problems of access and safety requiring the use of specialized access equipment, such as 'cherry pickers', or special scaffolding. Where access is difficult the use of two-part sealants may be an undesirable complication due to difficulties of mixing and large wastage due to the limited work life. One-part sealants are normally preferred for such work.

5.2 Cleaning and preparation of joint surfaces

5.2.1 Adhesion

Joints in buildings that are subject to movement can be sealed with gaskets, foams, tapes or with sealants.

Gaskets are preformed sections of synthetic rubber inserted into joints in a compressed state and designed to maintain a pressure against the sides of the joint to maintain a seal. No adhesion is needed between the gasket and the joint face. Gaskets can accommodate a limited degree of movement provided that they maintain an adequate degree of compression to sustain the lip seal pressure to seal the joint. The edges of the joint must be smooth and even to ensure a continuous seal. Expansive foams and tapes are also inserted into joints in a compressed state. There is, however, some adhesion to the joint faces although best results are obtainable if the edges of the joint are relatively smooth and even.

Sealants form a seal by bonding to both sides of the joint and forming a flexible link between the two substrates. Sealants accommodate movement by stretching and/or compressing. Maintaining a seal is dependent on the strength of the bond to the substrates being greater than the forces required to stretch or compress the sealant. Adhesion is therefore critical to ensure satisfactory performance. The basic requirements for adhesion are very simple:

1. Intimate contact between sealant and substrate (or primer layer)
2. Absence of weak layers or contamination at the interface, such as dirt and existing sealant residues.

Adhesion between sealant and substrate (or primer material) relies upon interfacial forces of attraction. It is therefore important that sealants and primers should be able to flow over and into the irregularities of the substrate surface, coming into contact with it so as to interact with its atomic and molecular forces.

Interfacial contact

Adhesion initially involves a liquid 'wetting' a solid surface. This means the formation of a thin film of liquid spreading uniformly over a surface without breaking into droplets (Figure 5.1b), for which the water break-free test is the simplest visual method of assessment. For wetting to occur, the surface tension of the liquid must be lower than the surface energy of the solid, resulting in a small contact angle θ. Hard solids such as metals, glass and ceramics, when clean, possess high surface energies which are readily wettable by sealants and primers. Hard plastics such as GRP have surface energies which are similar to the surface tensions of sealants and primers so that wetting is still achieved. However, softer plastics such as polypropylene and polyethylene have 'non-stick' low surface energies and wetting is sometimes a problem unless some form of surface modification is used.

Sealant and primer materials need to stick to the surface layers of materials; however, these layers can be very different to the substrate itself! With metals, this involves adhesion to metal oxides and with all materials the surface chemical groups are important. Weak, loosely bound layers and surface contamination (oils, grease, mould release agents, existing sealant residues and so on) must be removed to promote optimum adhesion. Surfaces can be made to be wettable and more 'receptive' towards sealants and primers by using pretreatments to alter the substrate's surface

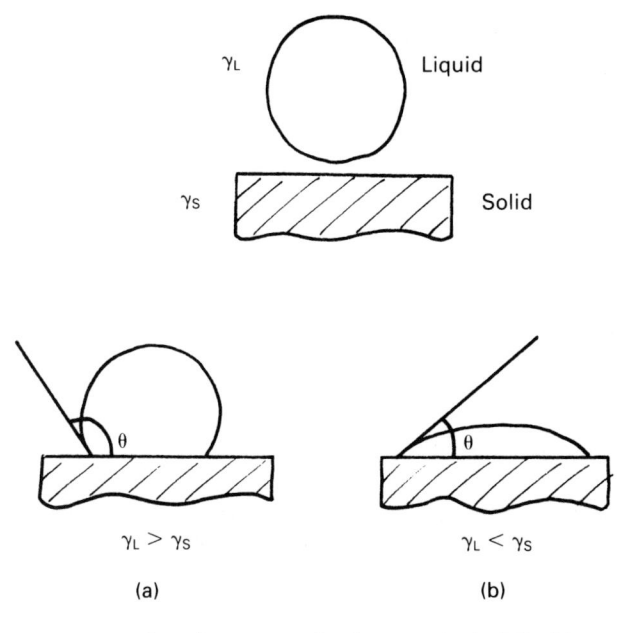

γ_L = surface free energy (surface tension) of liquid
γ_S = surface free energy of solid
θ = contact angle between liquid and solid

Figure 5.1 *Simple representation of wetting. (a) Non-wetting; (b) Wetting*

chemistry, energy and morphology (Sections 5.3 and 5.4). Naturally low viscosity primer materials are useful for binding friable surfaces such as concrete laitance, pre-wetting all surfaces, providing reproducible surface conditions for the sealant and providing water stability if coupling agents are used (Figure 5.2 and Section 5.7). In general, it is not necessary to achieve a high standard of surface preparation for short-term adhesion; however, a good standard of surface preparation is *essential* for maintaining long-term adhesion.

Mechanisms of adhesion

Having established interfacial contact, sealants and primers cure or harden. The forces acting across interfaces which prevent materials from separating under load are the subject of debate, and several mechanisms of adhesion have been proposed. However, the adsorption mechanism is generally

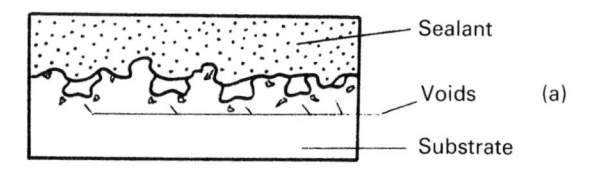

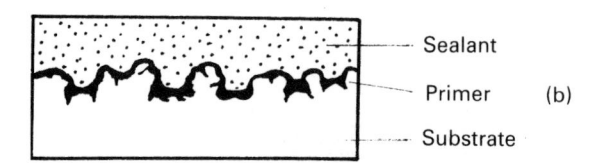

Figure 5.2 *Primers on a porous surface, showing also the principle of mechanical interlocking. (a) No primer; (b) use of primer to bind and consolidate surface*

favoured, with mechanical interlocking also playing an important role; diffusion bonding can also be important with some material combinations.

With the adsorption mechanism, sealant and primer macromolecules are physically adsorbed onto the substrate surface because of the natural forces of attraction between the materials. In effect, the polar nature of sealant and primer molecules acts like a weak magnet and they are attracted towards polar adherend surfaces such as glass, metals, polyester, PVC, ABS, acrylic, etc. The surface of polymer powder-coated aluminium, although smooth, is highly polar and this promotes bonding to sealants with large numbers of polar groups such as polyurethanes and acrylics. The most common interfacial forces are van der Waals forces, referred to as secondary bonds, although chemical bonding is involved in some cases. Chemical bonding can occur where there are reactive groups on the substrate surface which react with the sealant, or with constituents of it, or with certain primers and coupling agents.

Mechanical keying or interlocking of the sealant or primer material into the irregularities or pores of the substrate underlies the instinctive procedure of roughening surfaces to improve adhesion (Figure 5.2b). On porous surfaces such as concrete there is some penetration into the first few millimetres of substrate, providing an intermingling of primer, sealant and substrate. The penetration of primers into the microstructure of metal alloy oxide layers (such as anodized aluminium) may also be important, particularly in aiding retention of adhesion under wet conditions.

Diffusion bonding may occur where primer or sealant material penetrates the surface of the substrate. In some cases this occurs by a solvent effect of the sealant or primer on the substrate surface, causing softening and allowing the liquid to blend with the surface or surface layer(s). In effect, any discrete interface between the materials is destroyed. Such a situation arises, for example, where solvent-based acrylic sealant is applied to polyester powder-coated aluminium, or where primers are applied to existing sealant residues.

Primers

Primers assist adhesion by coating the substrate with a material designed to have a strong link with the substrate and presenting a surface that forms a chemical link with, or has a strong affinity for, the sealant.

Primers used on porous surfaces generally penetrate the surface, providing some mechanical link in addition to any natural affinity (Figure 5.2). In some cases the primer may also be required to bind and reinforce the surface, producing a firmer, stronger substrate to support the sealant and may act as a barrier to staining.

Primers for non-porous surfaces may be designed to form a chemical or physical link to the surface, or may produce a bond by diffusion. The exposed surface of the primer must be receptive to the sealant and may form a chemical or physical link (e.g. because of polarity).

The design of the primer is often unique to the application, having specific affinity to the sealant and to the substrate concerned. Substitution of primers can lead to catastrophic failure.

Where adhesion is formed by chemical or physical linkage, the timing of the sealant application may be critical as the reactive sites in the surface may be deactivated by contact with air, moisture or other contamination. Sealants should be applied within the recommended open time (see Figure 5.3) and care should be taken to ensure that prepared surfaces are not exposed to contamination.

5.2.2 Cleaning and preparation

Resealing may be necessary as a result of substantial joint failure, early minor defects or colour changes. The latter may require a special approach in order to avoid damaging satisfactory joint areas, limiting the methods of cleaning and preparation available. However, in such instances the nature of the existing sealing system is known, the causes of the localized failure normally fully appreciated and advice is available from the applicator and sealant manufacturer as to the best method of overcoming the difficulties. The

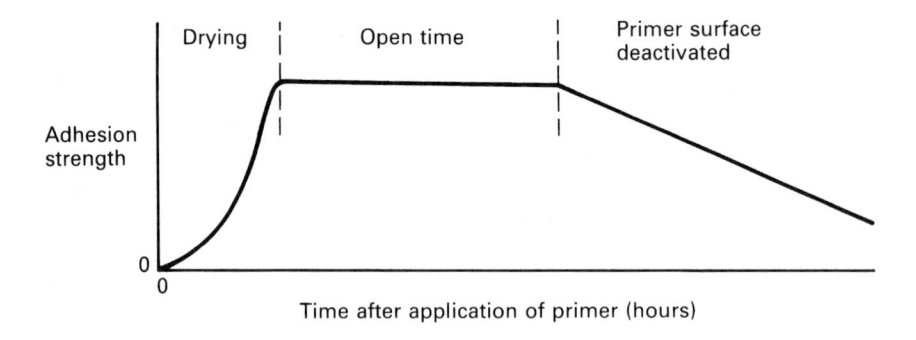

Figure 5.3 *Typical effect of open time on primers*

problems of removal of old sealant and preparing the joint to receive a new sealant are potentially more complicated.

In resealing older joints, failure may have occurred to a major or minor degree. In either case, it is probable that the exposed surfaces will be contaminated by dirt and debris carried in by wind and water. Where cohesive failure has occurred such contamination may be removed with the sealant, but where adhesion failure has occurred the contamination may have affected the surface of substrate forming the joint.

In the main there are two potential cleaning requirements:

1. *All* traces of the original sealant system should be removed to give, as far as is possible, a clean uncontaminated surface.
2. In cases where it is not possible to completely remove all traces of existing sealant, every effort should be made to ensure that *only the minimum film of residue is left*. Practically, or economically, it may not be possible to remove all traces of residue but it has been shown that *this will almost always be at the expense of performance*. A reduction in joint extension and overall joint performance is almost inevitable, in proportion to the type and amount of residue; the greater the amount of residue, the greater the reduction in potential performance. *It should be noted that non-curing and silicone sealant residues, particularly on porous surfaces such as concrete, have been found to severely reduce the performance of joints resealed with curing products.* However, identical products may well stick to thin residues of identical products, an option which may be available if it is suitable for other reasons to replace 'like with like'. It is recommended that thorough site trials should be undertaken in order to assess potential adhesion and compatibility. The final decision on cleaning requirements rests with the specifier.

Where substantial adhesion failure has occurred the removal of the remaining sealant from that surface is desirable as it is unwise to allow new sealant to bond to the old in view of the doubtful adhesion of that old material; all traces of the old sealant should therefore be removed before resealing.

It is always desirable to remove completely aged or failed materials and, in particular, non-curing products and silicones. However, where failure is totally cohesive, or totally concerned with the adhesion to one face of the joint with the sealant bonded securely to the other, consideration may be given to leaving a layer of sealant on the sound surface if the new sealant will bond to the old, provided that the modulus of the new material is less than that of the existing material. Joint preparation treatment however must not contaminate the surface of that existing sealant in an undesirable fashion and must not affect the adhesion in a detrimental way.

From the practical point of view the majority of sealant joint failures occur by loss of adhesion to one side or the other. In some cases adhesion failure may occur on both surfaces. Cohesive failure is a relatively rare occurrence but when it does occur special consideration should be given to the cause of that failure and the suitability of that sealant for the joint. Cohesive failure is frequently associated with excessive movement. Replacement of the joint seal with the same material may not therefore be satisfactory and replacement with an alternative seal may be necessary which in turn may necessitate total removal of the existing sealant material. Cohesive failure can also be caused by the use of an unsuitable sealant section or poor mixing of two-part materials.

Where consideration is given to replacement of existing sealant with new similar material, care should be taken to ensure compatibility by careful evaluation with suitable tests. Even where the new material is supplied by the same manufacturer and nominally the same grade, it is possible that formulation changes may have been made over the intervening years and there may therefore be differences between the new product and the old. With sealants having a long life expectancy it is probable that such changes will occur within that life time. Where the original manufacturer is not known it is likely that similar products from another manufacturer may be of sufficiently different formulation to be incompatible.

Some sealants will not adhere well to old cured material of the same type (*see* Appendix 2). In most cases therefore it is desirable to remove the existing sealant as far as possible in order to minimize the risks of incompatibility. Where it is decided that a layer of old sealant is acceptable and that the new material will bond to the old, it is preferable that the layer of old sealant shall be as thin as possible, as the new sealant will be softer and more flexible than the old; hence movement affecting the joint will tend to be concentrated in the body of the new sealant. It is essential therefore that the new sealant joint shall be as wide as possible. If the new sealant to be used is tougher than the old, total removal of the old sealant is essential, as otherwise movement will be accommodated primarily in the thin old sealant layer, increasing the stresses on the already aged and degraded sealant and hence increasing the probability of early failure.

5.3 Cleaning techniques

The techniques for cleaning joint surfaces on site are limited both by joint accessibility and by the nature of the materials forming the existing joint.

Most remedial work is carried out from cradles, ladders or scaffold, making the use of heavy equipment difficult if not impossible. Most of the equipment available is manual but angle grinders, power-driven wire brushes, mechanical chisels and grit-blasting techniques are possible in the right circumstances.

5.3.1 Bulk sealant removal

The majority of sealants are rubbery and plastic in nature and do not respond to mechanical removal techniques satisfactorily. If the sealant is present in quantity it tends to soften and spread rather than to abrade. This can spread contamination beyond the original joint and will frequently clog and choke the tools being used for its removal. It is essential, therefore, to remove the bulk of the sealant before using any mechanical aids. The primary means of removal is cutting and scraping using a variety of knives, wood chisels and similar implements (Figures 5.4 and 5.5). Lino knives and hacking knives are commonly used for the removal of the bulk of the sealant and, in some cases, special knives are fabricated to assist in the removal. Where the sealant is bonded to the back of the joint, special tools are frequently made in order to assist in cutting away from the inaccessible surface. After removing the bulk of the sealant, scraping with sharper knives will often enable the operator to reduce the contamination to a thin layer on the joint surface. Removal of the existing back-up material frequently permits easier access in order to reduce the contaminating sealant to a layer less than 0.5 mm thick.

If new sealant is to be applied to the old, the residual layer of sealant should be reduced to a very thin layer (or smear) but care should be taken to avoid snags or tears which may result in poorly bonded areas when the new sealant is applied (*see* Appendix 2). Where a layer of sealant is to be left in place, trials should be carried out to check whether solvent cleaning or priming affect the bond between new sealant and old, or the bond of the old

The reseal operation 69

Figure 5.4 *First cleaning stage – bulk sealant removal using hammer and chisel*

Figure 5.5 *First cleaning stage – bulk sealant removal using a knife*

sealant to the substrate. Any reduction in joint extension or unacceptable mode of failure arising from the trials should also be noted. The final specification should be determined in accordance with the results.

Where sealant is to be removed completely the contaminating layer of old sealant should be cut back to a minimum, cutting and scraping with sharp knives where practical to reduce that layer to the thinnest possible amount. Wetting the knives with water or an appropriate solvent as a lubricant may assist in some cases.

5.3.2 Further cleaning

When the contaminating layer has been reduced to the thinnest possible, further removal may be practical by either mechanical means or solvent cleaning. Mechanical removal is possible on surfaces that are not adversely affected by grinding or abrasion, such as stone and concrete (Figure 5.6). In some cases grit-blasting may be possible and practical tests will determine which type of grit is the most effective. Grit-blasting is usually carried out with carborundum or sand grits but in special cases plastic grit or egg shell materials can be used. Abrasive-blasting has also been carried out with volatile materials, such as 'dry-ice', which avoids the problem of removal of large volumes of abrasive material.

Figure 5.6 *Second cleaning stage – mechanical removal by grinding*

Grit-blasting is normally applied to both surfaces of the joint, but it is possible with larger sized joints to grit-blast one side whilst protecting the other side by a suitable masking device. Grit should be removed by vacuum extraction for safety reasons and to minimize dust retention in the joint.

Where grit-blasting is not practical, powered cutters or chisels may assist in the removal of sealant from the joint surface; alternatively, grinders with discs or abrasive wheels can be used. If too much sealant is left on the joint such abrasive wheels or discs will rapidly become clogged with sealant, reducing their effectiveness and spreading contamination. Frequent cleaning or replacement of the discs is often necessary.

Power-driven or hand-held wire brushes can also assist in the removal of the remaining traces of sealant, but care must be taken to ensure that they do not spread contamination or polish the sealant into the surface of the joint thereby creating a widespread, but less apparent, contamination.

5.3.3 Final cleaning

Solvent cleaning is usually the final stage of cleaning and preparation (Figure 5.7), but with modern materials and finishes it is often necessary to be cautious with the choice of solvent. Care should also be taken with regard to the original sealant materials in that some may be dissolved and

Figure 5.7 *Final cleaning stage – solvent cleaning*

spread by solvents causing staining or contamination even though the bulk may be removed. Bituminous sealants are particularly vulnerable in this respect.

The most widely used solvents are aromatic hydrocarbons such as toluene and xylene, and some chlorinated hydrocarbons. Toluene and xylene are aromatic hydrocarbons and have a good solvent power with a wide variety of polymers, oils and resins. Due to their high boiling points they linger long enough to allow materials to dissolve and soften before they evaporate so that the materials can be scraped or wiped off with a cloth. They are, however, flammable and harmful and should be used with great care and in accordance with the recommendations of the manufacturers and the Health and Safety Executive.

Chlorinated hydrocarbon solvents are very effective for removing oils and greases from a wide range of surfaces. Their high volatility and hence short drying time makes them useful on some surfaces where other solvents may adversely affect the surface. They are harmful and, although not flammable, the fumes react when heated and may generate very dangerous combustion products. On prolonged contact, some of these materials can also react with aluminium under certain conditions in the presence of water.

Both the aromatic hydrocarbons and the chlorinated hydrocarbons have been associated with adverse reactions of skin and soft tissue as a result of defatting. This can lead to dermatitis. The supplier's advice should be sought with regard to the usage of these products or suitable alternatives. Due consideration should also be given to the implications of the release of fumes on sites where there is an environmental policy with regard to halogenated products.

The use of solvent in conjunction with a mild abrasive, such as a nylon pan-scourer, will remove most sealant materials from smooth non-porous surfaces. Repeated cleaning can remove many materials from concrete, stone and other porous materials but it is difficult to remove deeply penetrating materials and to avoid staining.

Most solvents are potentially dangerous and should be used with care. Details of recommended handling procedures and the hazards associated with cleaning solvents are given in the Health and Safety Data sheets.

Abrasion is usually the last resort for removing traces of sealant from surfaces. It is not a viable alternative in many situations, such as cleaning metals, glass, and so on, but can, under the right circumstances, be acceptable with concrete, stone, brickwork, and some plastic materials. In order to achieve a clean surface it is essential that the bulk of the contamination be removed before abrasion, as otherwise the contaminant will tend to spread into the surface, clogging the tools, and continuing to spread contamination rather than removing it.

Grit-blasting can be used for final cleaning, especially on concrete and stone, as it reduces the risk of contamination being spread by cleaning solvents. Grinding tools and abrasive discs can also be used where the existing contamination has been reduced to a stain or to penetration level, but such tools are liable to spread contamination if there is any excessive amount of residue on the surface. Such tools also have practical limitations in that the grinding disc is several inches in diameter and cannot get into corners.

Mechanical wire brushes are not suited to final cleaning as they tend to leave a residue on the joint surface which requires solvent cleaning for its removal. Wire brushes can also polish surfaces which can result in reduced adhesion.

All abrasive techniques produce a considerable amount of dust which should be removed by vacuum or other means.

5.4 Suitability of cleaning method for surfaces

Modern methods of building construction use a wide variety of different materials. The physical and chemical characteristics of these materials impose limitations on the choice of methods and materials which can be used in the cleaning of surfaces.

In many cases joints occur between dissimilar materials. Where the two surfaces are too close to be cleaned separately, it is necessary to clean both with a technique suitable for the more vulnerable.

5.4.1 Concrete, stone and brickwork

Concrete, stone and brickwork can generally be cleaned by many methods including cutting and scraping, solvent cleaning and abrasives. The primary problem in most cases is represented by the irregularities of surface and the difficulties of cleaning into crevices and surface texture. Limitations also occur where these materials abut other parts of the building made with more delicate surfaces, such as painted metals or glass, which may limit the use of abrasives. Laboratory results have shown that non-curing sealant residues have a very detrimental effect on the adhesion of curing sealants used for resealing.

There are some treatments which can be used on masonry which can markedly affect the adhesion of sealants. These include treatment with waterproofing compounds such as organosilane oligomers used in consolidation of old stonework. Their effect on the performance of primers, and hence on the adhesion of sealant, will vary with time. Where these treatments are known or suspected, adhesion should be assessed by on-site testing at the time of resealing (Section 6.9.8).

Care is also necessary in the use of solvents, not only on account of more delicate surfaces adjacent to the joint but also of the nature of the sealant to be removed. Some sealant materials may be dissolved and carried into porous surfaces producing a permanent staining adjacent to the joint. Where such materials occur it is necessary to remove the contaminating sealant by mechanical means as far as possible, using solvent cleaning only to remove the last residual traces and using the solvent on a dampened rag to minimize the risk of permeation into the surface.

5.4.2 Metals and coated metals

Many of the metal windows in current use are finished with polyester powder coatings which impose severe limitations on the choice of cleaning techniques. Care is necessary in cutting and scraping and it may be necessary to use plastic scrapers rather than metal scrapers. Solvent cleaning can be carried out with care but many solvents affect the polyester powder finishes, changing the appearance if not literally dissolving the finish. White spirits and methylated spirits can be used safely on this type of finish but are poor cleaning and degreasing solvents. Chlorinated and aromatic hydrocarbons and most solvents should be used with great care as they can have a damaging effect on the finish, producing a change in appearance or even more severe effects. Solvent cleaning should be limited to wiping with a solvent-damped cloth and only undertaken after testing. The polyester finishes produced in the 1970s were noticeably more susceptible to solvent attack than the materials supplied currently. Sealants too can affect this type of finish in that solvent based acrylics will tend to dissolve and weld themselves to the surface of the polyester. Some primer systems will similarly become combined with the finish, whereas many of the sealants have true adhesion without any visible effect on the finish.

Anodized aluminium is more amenable to cleaning techniques because it is unaffected by solvents and relatively difficult to damage. Care must however be used when cleaning with metal scrapers, and plastic scrapers are generally more satisfactory. Nylon pan scourers are also generally safe and can frequently be used with solvents to remove tenacious residues.

Steel is generally protected by galvanizing or similar anti-corrosion treatment and any cleaning procedure must take this into account. Abrasive treatments must be used with great care to minimize damage to the anti-corrosion coating. Paint finishes may be either traditional paints, polyester powder coatings or other types. Most paints are susceptible to chlorinated solvents and aromatic hydrocarbons, both of which should be used with care. White spirit and methylated spirits are less likely to affect the finish but ketones and ester solvents may affect polyesters and some paints.

5.4.3 Timber

Timber is generally finished with paint, translucent microporous finishes or some preservative finish. A few hardwoods may be left untreated but, if left untreated, the surface tends to become highly absorbent and fibrous in nature. Paint finishes are susceptible to solvents and abrasives and should be cleaned with care. In many cases the adhesion of the paint finish to the timber is unreliable and, in order to ensure a safe anchorage for the resealing operation, it is often desirable to remove the existing timber finish in preparation for priming and sealing with materials known to adhere satisfactorily to the timber substrate.

Microporous finishes tend to erode on the exposed surfaces but are more durable where they are protected from sunlight and weather. Again adhesion tends to vary and in general it is better to ensure that the surface finish is removed to ensure sound adhesion.

Some preservative timber finishes may incorporate waxes, oils, silicones, and so on which can have a detrimental effect on the adhesion of sealants and other materials. Where transparent penetrating finishes have been applied, it is generally advisable to check the adhesion and cleaning and priming procedures by on-site tests prior to specifying finally the cleaning and resealing procedure.

In the case of some finishes which contain substantial quantities of wax it may be necessary to remove the surface of the timber by planing or routing to achieve a satisfactory surface to receive the sealant. Where timbers have been treated with slow-drying oils it may be necessary to solvent clean thoroughly using aromatic or chlorinated hydrocarbon solvents to remove the bulk of the oil, and to allow a period of weathering to dry the residual oil materials in order to achieve a surface to which the sealant can stick.

5.5 The failed sealant

The method of cleaning may also be dictated to some extent by the nature of the failed joint sealant and its condition at the time that resealing is to be carried out.

Most sealants are rubbery/plastic materials, somewhat thermoplastic in nature and if subject to abrasion they tend to spread, clogging the abrading tools and will tend to melt and flow into the joint surfaces spreading contamination. It is therefore important to ensure that most of the sealant is removed before any abrasion is attempted.

5.5.1 Non-curing sealant residues

Laboratory work has demonstrated that it is essential to remove all traces of non-curing residues in order to realize adequate performance from a resealed joint (*see* Appendix 2). Sealants such as oil-based mastics, bitumen/pitch, butyl and acrylic-based materials are readily soluble in a variety of solvents and they tend to spread and contaminate surfaces if subject to solvent cleaning. It is important therefore that these materials should be substantially removed without solvents to minimize risks of the spread of contamination, and that subsequently the surface should be cleaned two or three times with suitable solvents to remove the contamination; clean cloths should be used on each occasion in order to remove the contamination prior to resealing.

Oil-based mastics

The method of removing oil-based mastic depends on its condition at the time the cleaning is attempted. Aged oil-based mastic tends to be hard, brittle and substantially insoluble, and can be removed by mechanical means with solvent cleaning as the final treatment. Oil-based mastics in the soft live condition are soluble in a wide range of solvents. These solvents are liable to spread contamination if applied to joints where there is a substantial volume of mastic in place. It is therefore important that most of the mastic is removed by scraping, cutting, wiping, and so on. Solvent cleaning should be restricted to the final stages and carried out in two or three operations, wiping off the solvated sealant between operations and ensuring that the final cleaning is carried out with clean cloths.

Bitumen-based sealants

Bitumen/pitch-based sealants are readily soluble in hydrocarbon and other solvents, and can spread contamination and staining to a very considerable degree. It is again important that bitumen-based sealants should be substantially removed by cutting and scraping prior to the use of solvent cleaning. In many cases repeated cleaning with cloths moistened with solvent is the only practical way to remove the residue, taking care to avoid spreading the contamination and ensuring that the solvated bitumen is removed at each cleaning operation. Clean cloths are essential for the final cleaning. Where possible, emulsifying solvent cleaners (mixtures of solvent and oil-soluble detergent) can be used for removal of bitumen residues and the emulsified residue washed off thoroughly with water to minimize contamination. This type of treatment is not often practical unless there is a secondary seal in the joint which is not susceptible to solvent attack.

Butyl-based mastics

Butyl-based mastics remain soft and sticky for a very long time. Even when the surface is dried they are frequently soft and sticky in the mass. They are soluble in hydrocarbon and other solvents, spreading a sticky adhesive contamination. It is important that the seal should be removed as far as possible by cleaning and scraping before using solvents. Solvent cleaning should then be carried out in two or three stages using solvent-damped cloths rather than a liquid solvent, ensuring that the final cleaning operation is carried out with clean cloths. Alternatively, emulsifiying solvents can be used and washed off with water where this method of cleaning is acceptable.

Acrylic sealants

Acrylic sealants are extremely difficult to remove. They are tough, thermoplastic, and very adhesive. Acrylic sealants are soluble in aromatic hydrocarbons, chlorinated hydrocarbons, and some other solvents. Solvents tend to dissolve the polymer from the surface of the sealant, spreading contamination rather than removing the bulk. It is important that the mass of the sealant should be removed by careful cutting and scraping followed by repeated solvent cleaning using cloths moistened with solvent. Any attempt at abrasion is liable to be unsatisfactory as the acrylic resins will melt, flow and clog the grinding tools and spread the contamination in the joint.

5.5.2 Cured sealant residues

Polysulphide and polyurethane sealants

Laboratory trials carried out during the RESEAL project showed that the presence of small amounts of polysulphide and polyurethane residues were the least detrimental to the performance of joints resealed with curing materials (*see* Appendix 2). Polysulphide and polyurethane sealants tend to toughen with age and, as a result, aged sealant can be removed more readily by abrasion than freshly applied material. These materials tend to cut and tear so that a large proportion of the sealant can be removed by cutting and scraping techniques. The residue can then be removed, either by abrading off the bulk and then solvent cleaning, or by the use of solvent and mild abrasives, such as nylon pan scourers, to remove the residue from non-porous surfaces. This technique requires larger quantities of solvent to be applied and great care should be taken because of the associated hazards. Scraping, abrasion and solvent cleaning can all be used with weathered materials but abrasion is less satisfactory with freshly applied materials.

Where two-part materials have been used, problems may occur due to poor mixing and poorly mixed pockets of uncured fluid sealant may be encountered. Such material is easily spread onto adjacent surfaces causing staining and contamination. Special care is advisable in such situations, with particular attention to removing the bulk of the sealant by cutting and scraping to minimize contamination.

Silicone sealants

Silicone sealants can be substantially removed by scraping and cutting, and by moderate abrasion. Wire brushes can be used on masonry surfaces, or slow drying solvents with pan-scourers and similar mild abrasives on non-porous surfaces. However, surface contamination can be left which is difficult to remove and which may interfere with the application of other sealants. It is difficult to remove silicone residues sufficiently well to ensure good adhesion with sealants other than silicones, but a thin layer or smear of residue will be less detrimental than a thicker one.

On porous surfaces the contamination is unlikely to interfere with adhesion if primers are used. On non-porous surfaces contamination may give rise to problems, but most can be removed by solvent cleaning or by the use of silicone digesters and washed off with water. Heavy contamination can be removed by the use of digesters in paste form by repeated applications. Silicone contamination is difficult to see on most surfaces although it is frequently visible as a shadow line on light coloured masonry or stone. This can sometimes be detected by applying a fine mist spray with water, contaminated areas being more resistant to wetting with the water forming discrete drops. Where it is impossible to remove in depth, resealing with compatible sealant and primer is generally to be advised.

5.6 Final preparation

The removal of the old sealant and cleaning of the joint surfaces tends to be a messy job, and is usually best carried out prior to the main resealing operation. This reduces the risk of accidental contamination and damage to the resealed joints. However, as a consequence of the delay, the joints may become contaminated by dust or other site contamination which should be removed immediately prior to priming and sealing. Dust should be removed by brushing the joint with a clean soft brush or wiping with a clean lint-free cloth. Other contamination may be removed by wiping with a cloth damped with suitable solvent, followed by wiping with a clean dry cloth.

5.7 Priming

Primers are not always recommended with sealants but *adhesion failure is often due to lack of priming*. Primers are used to improve the adhesion of the sealant to the substrate. That improvement may result in increased adhesive strength and/or increased bond durability. Primers function in various ways depending on the nature of the surface and the sealant to be joined.

Primers for porous surfaces are generally designed to penetrate into that surface, forming a mechanical interlock as well as an adhesive bond. The primer may also be designed to seal the surface, reducing its porosity. Where surfaces are weak or friable, the primer can penetrate and bind that surface, increasing the strength of the surface and hence improving the adhesion (*see* Section 3, Figure 3.1). Such primers are usually film-forming and based on resins and polymers that provide a good surface to receive the sealant. Primers for porous surfaces such as concrete or masonry may also be designed to adhere to damp surfaces, either by the use of special surfactants or by the use of moisture-reactive materials such as polyurethane resins.

Adhesion to non-porous polymeric surfaces such as coated aluminium may be improved by primers that penetrate into the surface by swelling or dissolving the surface creating a mechanical link, or by the use of chemical primers that change the nature of the surface and present a more compatible surface to the sealant. Most of these chemical primers are based on silane coupling agents, applied as a dilute solution. Ideally this results in the deposition of a single layer of molecules on the surface, one end of the silane molecule having a strong affinity for the surface and the other having a strong affinity for the sealant. These primers are sometimes described as surface conditioners.

Many of the modern primers are reactive materials, creating a chemical link between sealant and primer. The reactive groups in the primer surface that create these chemical bonds will also react with moisture and other atmospheric materials. If the sealant is not applied within the recommended time the primer surface gradually becomes inert. The fully reacted or cured primer surface may not be receptive to the sealant, giving a weak bond. It is important to ensure that the sealant is applied within the recommended time to ensure satisfactory adhesion (see Figure 5.3).

In resealing joints priming may not always be necessary, but most cleaning techniques will damage or remove the existing primer. Chemically reactive primers tend to become insoluble and more difficult to remove but, since the cured primer no longer offers a sealant-receptive surface, re-priming is usually necessary. When resealing joints, priming will sometimes improve results even where no primer was originally needed. Primer may reduce the effect of joint contamination and it may improve the adhesion of new sealant to traces of the old material.

On-site selection and testing of sealants and primers is not easy but testing on site-prepared surfaces is essential in preparing the remedial specification.

Primers are particularly important in improving and maintaining adhesion under prolonged damp and wet conditions which may not be easy to reproduce under site conditions. It is often easier to remove specimens from the site in order to carry out water immersion and other simulated weathering tests in the laboratory.

The application of primers requires special care (Figure 5.8). Most primers are designed to be concealed by the sealant and their appearance, colour and weathering properties are not primary considerations. Many discolour on weathering and some degrade and erode when exposed. Most will change the appearance of the surface to which they are applied and some will stain. Special care is needed to place the primer where it is needed but not elsewhere. Masking joints with masking tape is not a good solution as many primers will penetrate under the tape, staining the surface and in some cases sticking the tape firmly in place.

Careful application of primer is essential to ensure that it is confined to the joint area, particularly with visible joints. Primers should preferably be applied before back-up foam is inserted to ensure good adhesion over the full depth of the joint. It should be ensured that primer is not applied to the surface of the foam, which could result in adhesion to the back of the joint causing three-sided adhesion. Primers should be allowed to dry before inserting back-up foam as the oversize foam will wipe the primer from the joint face.

Where masking tape is used it should be applied after priming, taking care to ensure it is not applied to wet primer as this could cause strong adhesion of the tape and make removal difficult. Primers should be applied and

Figure 5.8 *Priming*

allowed to dry in accordance with the manufacturer's instructions. Premature application of sealant can result in reduced adhesion and may interfere with chemical cure of some sealants.

Primed surfaces should be protected against dust and contamination. Protection against rain and water before sealant is applied is desirable, particularly with primers that react with moisture. Excess moisture can reduce the open time of the primer, causing a reduction in adhesion of sealant applied later.

Sealant should be applied within the open time period recommended by the manufacturer to ensure good adhesion between sealant and primer (*see* Figure 5.3). Where this is not possible, the surface should be reprimed.

5.8 Joint depth

The depth of sealant in the joint is important, both in the interests of performance and economy. Fully elastic sealants confer optimum movement accommodation when used in thin sections whereas plastic sealants perform best when the depth (D) is at least equal to the width (W). Elasto-plastic materials tend to give the best results when the depth is half the width.

A minimum depth of sealant is necessary both for durability and for ease of application. Sealants are applied under a wide variety of conditions, and the joints to which they are applied are formed in various ways and under varied conditions. There must therefore be a reasonable degree of tolerance on the width and depth of the joint seal to accommodate these conditions.

For normal applications in cladding and similar circumstances where the joints are not subject to traffic or prolonged pressures, the recommended seal depths (Figure 5.9) are:

Sealant type	Min depth	Normal guideline	Max depth
Plastic sealants	6 mm	D=W	25 mm
Elastoplastic	6 mm	D=W/2	50 mm
Elastic sealants	6 mm	D=W/2 or W/3	20 mm

Where joints are subject to pressure or traffic, the depth of sealant is normally increased to give better resistance to pressure on the joint. The increased depth increases the stress when the joint is affected by movement, resulting in some reduction in the movement accommodation capability of the joint. In these deeper joints the Movement Accommodation Factor may be reduced by 10% or more, increasing the joint width required to accommodate the joint movement.

5.8.1 Joint depth control

Joint depth is controlled by the use of a back-up foam or other joint filler material (Figure 5.10). Where joints are subject to movement, it is important that the sealant is bonded to the joint faces only and that it is not bonded to

Plastic sealants — Depth = width

Elasto-plastic sealants — Depth $= \dfrac{\text{width}}{2}$

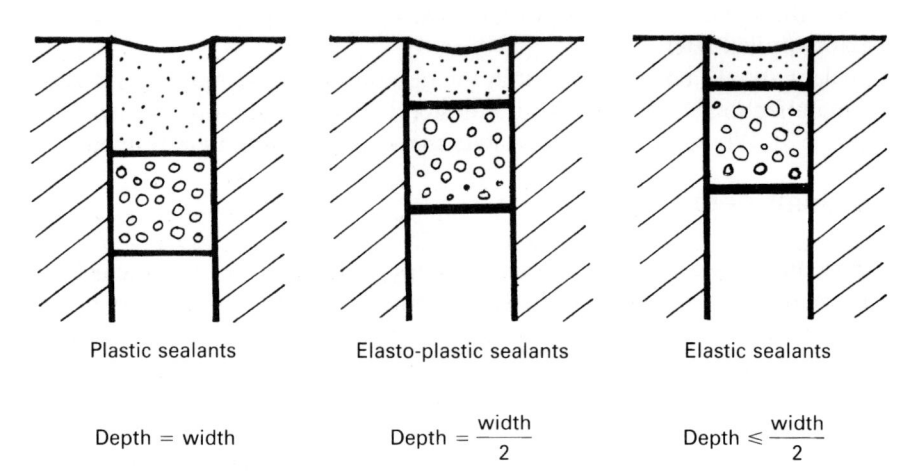

Elastic sealants — Depth $\leq \dfrac{\text{width}}{2}$

Figure 5.9 *Joint depths for sealant types*

Figure 5.10 *Joint depth control – insertion of backing material*

the back of the joint. This enables the sealant to stretch or compress without impediment. The joint back-up or filler must therefore be a material to which the sealant will not stick, or it must be covered by a bond-breaker tape to prevent the sealant sticking.

Cellular foam materials are preferred as back-up for sealants because they can be compressed in the joint initially, and will expand when the joint opens and compress when it closes, thus following the joint movement and providing continuous support for the sealant – particularly during the cure period when the seal is easily deformed or distorted. Open-cell materials tend to absorb water and can act as wicks, drawing water up behind the seal and creating dampness problems. Closed-cell materials are preferred as they do not absorb water and do not spread dampness. Closed-cell expanded rubbers are suitable but expensive. Expanded closed-cell polyethylene foam is particularly suitable as it is non-absorbent; most sealants do not stick to it, and polyethylene is inherently hydrophobic and tends to repel water.

Closed-cell back-up foam can cause blistering in joints subject to large cyclic movement. If the foam is damaged during processing, handling or as it is inserted in the joint, groups of cells are torn or ruptured. When the foam is compressed as the joint closes, the air in these damaged cells is forced out and air drawn back in as the joint opens. If the torn surface is covered by plastic or partially cured sealant, the air blows a blister in the sealant. When the joint re-opens, the torn surface of the foam closes and air is drawn back into the cells from elsewhere. The blisters gradually get larger with further cycles of movement. This problem is most likely to occur in joints between well insulated panels of aluminium or plastics. Most damage to the foam occurs as the foam is compressed into the joint, especially if rough or sharp tools are used. The use of joint rollers to insert the foam will reduce the risk of damage and minimize blistering; it does not occur where open cell backing foam is used.

Joint filler materials, such as polystyrene foam or fibreboard, are not good back-up materials for movement joints because, although they will compress to allow closing movement, they do not expand to follow opening movement of the joint and therefore leave the seal unsupported. Many sealants adhere to these materials unless the surface is covered by a bond-breaker tape. Polystyrene foam is attacked by many solvents that are used in sealants and primers. Fibreboard joint fillers may be impregnated with bitumen or other materials which are easily extracted by solvents, spreading contamination and may also cause staining of sealants if not isolated adequately.

5.8.2 Back-up materials

The most widely used joint sealant back-up is polyethylene foam. This material is available in various grades and forms. It is supplied as cut sections, sheet or in rod-form (Figure 5.11).

Rod-form polyethylene foam is preferred as it has a thin surface skin which ensures minimum adhesion between sealant and back-up. It also has advantages in that it is difficult to twist or insert the wrong way. Provided it is used at least 25% oversize it results in the seal having a large bonding surface with a slightly waisted section, giving good stress distribution when deformed by movement. Rod-form back-up has disadvantages in that if used in sections less than 20% oversize the difference in seal depth across the joint can cause excessive strain in the seal; also the large round section of back-up requires a deep joint to accommodate it. Rod-form foam tends to be expensive.

Cut foam, either sheet or section, has a surface of cut open cells into which the fluid sealant will penetrate. With large cells this causes an interlocking effect which impedes the separation of sealant from back-up and reduces the ability of the joint to accommodate movement. Material with small cells separates more easily and does not interfere with seal performance. Only cut foam with small cells should be used in joints subject to substantial movement. Some polyethylene foam sheet and cut sections are available

The reseal operation 85

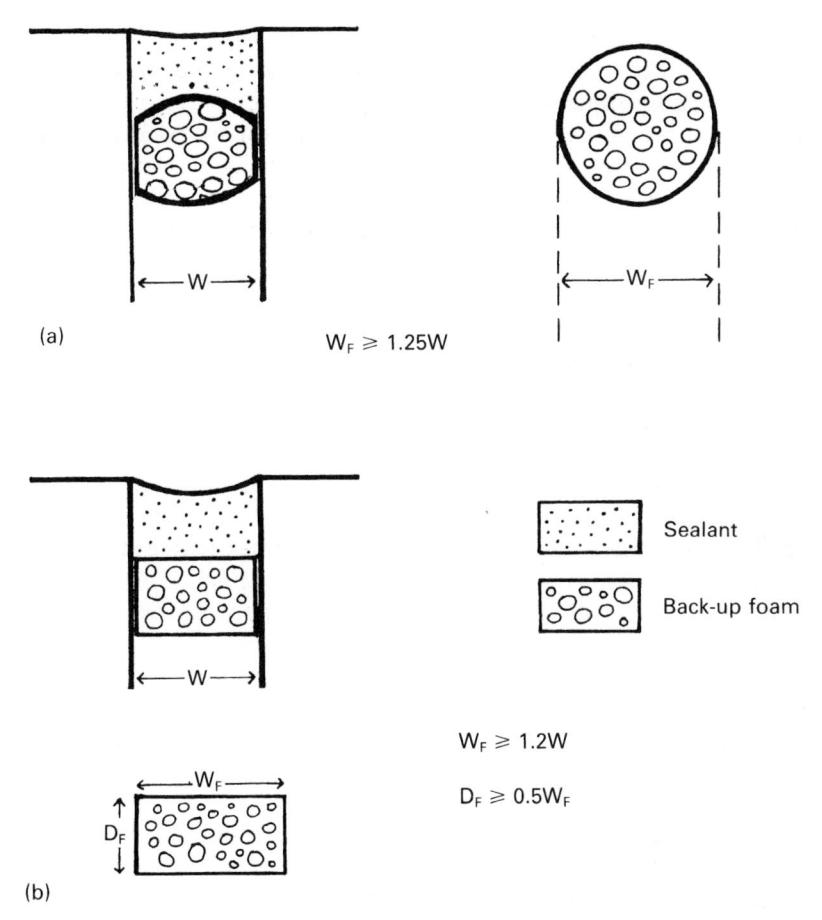

Figure 5.11 Recommended back-up foam sections. (a) Rod form, round section; (b) rectangular cut foam sections

(a) $W_F \geq 1.25W$

(b) $W_F \geq 1.2W$
$D_F \geq 0.5W_F$

with a skin on one surface; this is ideal provided that the skin is on the surface against the sealant as this ensures minimum adhesion and easy release of the sealant. When the sealant has cured, the back-up foam should release from the sealant, allowing the seal to deform freely.

Cut sections of foam should be at least 20% wider than the joint to ensure the foam is under sufficient compression to expand and follow joint movement during the cure period. The section should be at least half as deep as the width to minimize buckling when it is compressed in the joint (*see* Figure 5.11b). Cut sections of foam can easily be distorted or twisted as they are inserted in the joint, causing uneven seal sections and consequential weaknesses that may cause premature failure. The use of thin sections of foam can cause buckling and consequential bulges or hollows in the seal, giving an untidy appearance.

Insufficient compression of the foam can result in a loose backing which does not support the sealant as it is applied, resulting in poor compaction and bulging or slump after tooling. Insufficient compression of the foam may also cause the back-up to become loose as the joint opens, placing stress on the corners and causing splits or tears in the back of the seal which may progress to cause premature failure. Cut polyethylene foam is relatively cheap and readily available to any size required.

Sponge or expanded rubber (EPDM or butyl) can also be used as a joint backing and is particularly suitable for load-bearing joints because it can provide physical support to the joint seal without impairing movement accommodation. Care should be taken to ensure that the sealant does not adhere to the rubber, which should be covered by a bond-breaker tape if necessary. Rubbers, both natural and synthetic, contain curing agents, antioxidants and extenders that can migrate into sealants causing staining or discoloration. Where light coloured sealants are used, it may be necessary to separate the seal from the rubber by a section of polyethylene foam or backing strip.

Polyurethane foam is not normally used as a sealant back-up because most sealants adhere to the foam. It is also unsuitable as it is an open-cell sponge and acts as a wick, drawing water up behind the joint wherever there is access. A special form of polyurethane foam has been used for sealing joints between plastic and aluminium panels where large rapid cycles of movement give rise to blistering problems with polyethylene. The open-cell foam does not cause blisters and by using a very low tear strength foam, sealant separation is achieved by tearing the surface from the foam. A modified form of polyethylene foam, reputed to be open-cell along its length, is available in the USA which is claimed to overcome the problem.

Care should always be taken to ensure that the backing foam is not stretched as it is fitted, and left in tension, as the foam will gradually recover and can displace the sealant producing thin sections or gaps.

Special sections of foam or very soft foam may be necessary when sealing joints having an irregular or non-uniform section.

5.8.3 Bond-breaker tapes

Self-adhesive strips of polyethylene or PTFE are suitable bond-breaker tapes for use with most sealants. They have smooth surfaces from which the cured sealants can de-bond easily, allowing the sealant to deform as the joint moves.

Other materials, such as PVC insulating tape or paper masking tape can be used with some sealants, but many sealants can develop a strong bond to these materials rendering them ineffective. Thin self-adhesive strips of polyethylene foam are also used as bond-breakers, particularly for sealing joints which are not deep enough to accommodate a thick section of foam back-up but too deep for a thin polyethylene or PTFE tape.

Bond-breaker tapes are useful when sealing stepped joints, joints over filler boards where the board is not set back sufficiently to accommodate a normal foam back-up, or where joints are subject to traffic and need a firm support. Bond-breaker tapes are also used in overseeing situations to prevent adhesion of new sealant to old.

5.9 Mixing of multi-component sealants

5.9.1 Sealant materials

Most building sealants are single-pack materials, which set or cure either by solvent loss or by chemical reaction initiated by contact with atmospheric moisture or oxygen. Most single pack polysulphides require a relative humidity in excess of 35% to cure, whereas most silicone sealants will cure at 20% RH or less. Such sealants tend to set or cure slowly from the surface inwards, the centre curing more slowly than the exposed surface.

Some sealants are supplied as multi-part materials having two or more separate components that must be mixed together immediately prior to application. Polysulphide, polyurethane, silicone, epoxy and some other sealants are available in this form. In some cases the multi-component sealant is used because it is faster curing than similar one-part materials, but in other cases the product is not available in a single pack form. When properly mixed, multi-part materials cure homogeneously with consistent properties throughout the mass of sealant. The components contain the base material and the curing agents in separate parts, and in some cases colouring pastes may be included. The individual components are stable separately but the curing action is initiated when they are mixed together.

In most cases the sealants are supplied with the individual components packed in separate containers in the correct proportions, but some gun-grades are available packed in layers in a single container in the correct proportion so that they only need to be stirred to mix. Where materials are supplied in separately proportioned containers, it is essential that the entire contents of these containers are transferred into the mix; any deficiency may affect the rate and degree of cure achieved and may also affect the colour. Mixing of these multi-component products is critical in order to ensure even distribution of the curing agents and other materials throughout the mass of sealing compound.

In some cases the different components are different colours, which assists the operator to see the mixing action, especially in the early stages, but the colour differences become less obvious as mixing proceeds. However, it is not always possible to achieve a substantial colour difference between components and in some cases pigmentation of reactive curing agents may cause instability. Although there may be distinct colour differences in some products, there are others in which there is little difference and even some in which the curing agent is a clear liquid which rapidly disappears in the mass of sealant. The disappearance of streakiness is not a reliable indicator of thorough mixing. It is essential that the applicator should develop a technique that ensures thorough and even mixing of the components.

Even distribution of curing agents is essential in order to achieve the design properties of the sealant. Most manufacturers allow some tolerance on mix proportions where possible. In some sealants, a modest deviation in mixing proportions is acceptable and has little effect on properties or cure time. With other sealants, however, a slight deficiency can result in a soft sticky mass, whilst an excess may cause overcure and a tougher shorter product. Thorough and even mixing is therefore essential.

5.9.2 Mixing techniques

Most multi-part sealants are supplied in cans or other containers with the base-component can having sufficient ullage to allow the other parts to be added and the sealant mixed in the can. Mixing may be carried out by hand, by bench mixer or by hand-held drill with a suitable mixing blade.

The majority of site-applied multi-part sealants are mixed by slow speed drills fitted with a mixer blade. This technique adapts easily to the variety of container sizes used and suitable drills are available for building site and domestic power supplies. It is generally recommended that the drill speed should not exceed 300 r.p.m. because excessive speed causes heating and aeration, resulting in reduced application life and sealants having a spongy consistency. Aerated sealants may have a reduced performance and durability.

Mixer blades also come in various shapes but the most common are the horse-shoe or the spiral (see Figure 5.12); T-bar and triangular mixer blades are also used. They are generally less effective than the horse-shoe or spiral, but may have advantages when working with very viscous materials because they reduce the load on the drill, but require longer mixing times.

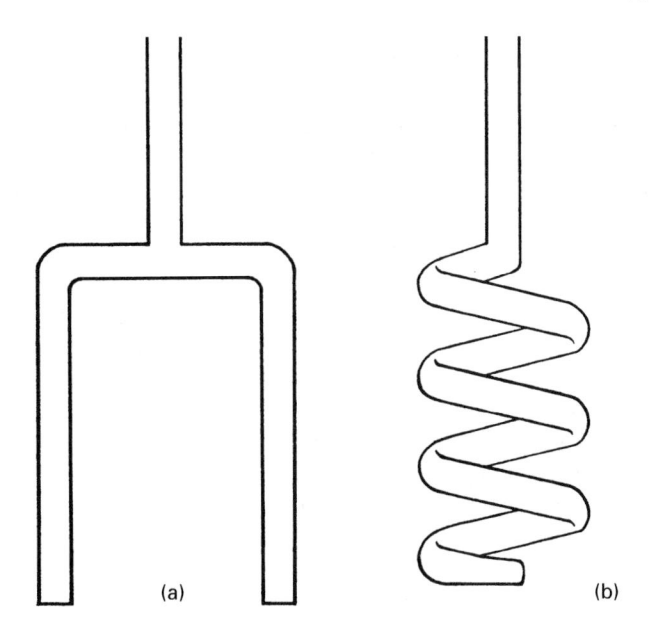

Figure 5.12 *Mixer blades. (a) Horse-shoe blade; (b) spiral blade*

The horse-shoe blade is an efficient stirrer but produces little up and down movement in the mix. It is therefore important that the operator moves the mixing blade up and down in the mix to achieve thorough mixing and even distribution. The spiral blade is designed to push the sealant down towards the bottom of the can in order to achieve a down and up circulation as well as the rotary action of the drill. It is a very efficient mixer but, if allowed to come above the top surface of the mix, it tends to entrain air.

Both mixer blades are smaller than the diameter of the can and must therefore be moved round, up and down and sideways to ensure that the entire contents of the can are thoroughly mixed, while, at the same time taking care to avoid entraining air. The rotating blade should be used to scrape the sides and bottom of the can, with particular attention to the corners around the base. It is normally necessary to stop during mixing and scrape down the sides of the can to ensure that all the curing agent is in the mix and not up the side of the can. Two litres of gun-grade sealant can be mixed in about six minutes but the time will vary with different operators and materials. Mixing time is less critical than the stirring action in ensuring a thorough and even mix. The spiral blade is probably the more efficient mixer but is more difficult to clean than the horse-shoe. Both designs will produce a satisfactory mix with most sealants.

90 Resealing of Buildings

Bench mixers are used primarily in factory situations. They are usually restricted to one or two can sizes and tend to be permanently sited and thus less versatile than the drill and paddle for site work.

After mixing, gun grade sealants are usually transferred into the barrel of a bulk loading gun or into cartridges for use in other types of gun. After the mixer blade has been removed, the sealant surface should be smoothed to avoid entraining air. A plunger plate can then be placed on top of the sealant. A cartridge or barrel gun should be placed over the hole in the plate and the gun/cartridge and plate pressed down onto the sealant (*see* Figure 5.13), forcing the sealant through the hole and into the cartridge/gun barrel until it is almost full. The cartridge/gun barrel should be lifted and twisted off the plate taking care not to drip sealant down the side. The plunger can then be inserted in the cartridge or the cap and nozzle fitted to the gun ready for use.

Multi-part sealants have a limited application life after mixing and it is essential that they should be used within that time. They may otherwise

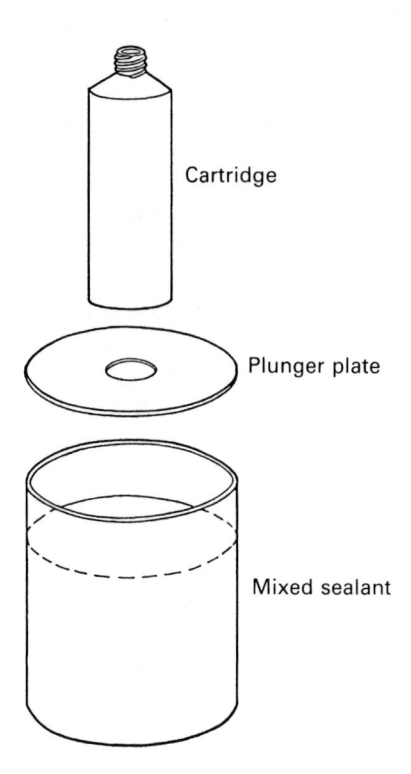

Figure 5.13 *Cartridge filling*

have achieved a partial cure, making them difficult to extrude, and reducing their ability to wet surfaces so causing poor adhesion.

High temperatures will shorten the application life of the product and may cause slump on application. Conversely, low temperatures may extend the work life and cure time; an increase in viscosity of the material may also lead to poor or incomplete mixing. Some sealants generate heat as they cure so that large masses of mixed sealant tend to get hot, accelerating the cure and shortening the work life. It is advisable to transfer bulk sealant into cartridges or other small containers immediately after mixing to minimize this effect.

5.10 Gun application

Most single-part sealants are applied by gun. Some flowing grades are poured into wide, upward-facing joints, but the majority of joints are sealed by gun-applied materials.

Most sealants are supplied in cartridges or flexible sachets that are loaded into a suitable gun and extruded (Figure 5.14), with the applicator disposing of the pack in an appropriate manner when empty. Some barrel guns may be filled directly into the barrel from bulk sealant or mixed multi-part sealant. Multi-part sealants must be applied within their application life after mixing.

Figure 5.14 *Gunning sealant*

Various types of sealant guns are available from sealant manufacturers and other sources. The most popular types are hand operated, using a pivoted trigger to force a piston down the barrel of the gun or the cartridge, forcing the sealant through a nozzle into the joint.

Cartridge guns are available with full barrels, half barrels and side bars only. The side bar or skeleton gun is the lightest, easy to load and probably the most common. The half-barrel gun was easy to load, but less reliable and easily damaged and is now less common. Full barrel or closed barrel cartridge guns are less common but preferred by some professional applicators. Choice is mainly a matter of personal preference.

Barrel guns are used for bulk loading and for flexible sachets. Bulk loading is now less common as it involves more cleaning at the end of the day and, particularly, solvent cleaning with its associated health and safety hazards. The flexible sachets reduce or eliminate the cleaning difficulties and reduce disposal problems. Barrel guns tend to be preferred by professional applicators because of their robust construction and the variety of nozzles available.

Air-operated guns are available to suit cartridge packs, flexible sachets and bulk loading. Air-operated guns are seldom used on building sites because of the need for compressed air and the restrictions imposed by air lines, compressors, etc. Air-operated guns are also more difficult to control with the varying sizes of site joints and variation in accessibility, whereas the hand gun is very easily controlled by varying the hand pressure and rate of operation. Air-operated guns offer advantages where there are long runs of uniform joint section to be sealed with good access to the joints.

Pumping of sealants from a bulk supply is very rarely used on site. The problems of pumping gun-grade sealants through flexible hoses are likely to limit the use of this technique to factory and workshop applications.

Sealant cartridges may have an integral nozzle, or a screw thread at the top to allow separate nozzles to be fitted. Various shapes and sizes of nozzle are available made from metal (diecast alloy) or plastics (polyethylene, polypropylene or high density polyethylene), some of which can be reshaped by heat and pressure where special nozzles are needed.

Flexible sachets are used in barrel guns similar to those used for bulk loading. These guns have a variety of nozzles available that screw into the end cap of the gun. These nozzles are made from metal or plastics.

Plastic nozzles can be cut to size with a sharp knife and most can also be shaped by softening in boiling water and holding in the shape desired until cool. Plastic nozzles do not scratch or damage surfaces easily but they are quickly worn when used against rough surfaces such as brick, stone or concrete.

Metal nozzles are available in a range of sizes and shapes to suit each type of extrusion gun. They are not as easily modified as plastic ones but some special shapes can be produced. Metal nozzles can damage painted or anodized surfaces. Although they do not wear as quickly as plastic when used against brick or stone, they can develop sharp edges that easily damage paint or other finishes.

The nozzle should be of a size to suit the width of the joint to be sealed. When sealing butt joints the end of the nozzle should be slightly smaller than the joint so that it fits into the joint slot. This should enable sealant to be firmly extruded into the joint without overflowing onto the faces of the panels either side. When applying fillet joints the nozzle should be cut to the face width of the fillet. The nozzle end is normally cut at an angle of 45° so that the sealant is pushed down into the joint as it is extruded and flow is controlled by the tip of the nozzle.

The extrusion of the sealant is achieved by pulling the trigger of the gun. The rate of extrusion must be matched to the speed of moving the nozzle along the joint to ensure that sufficient sealant is applied to fill the joint to the required depth (*see* Figure 5.15a). Too little sealant can result in a shallow joint with a consequential reduction in performance, durability or load bearing (Figure 5.15b). Too much sealant will result in overflowing the joint and consequential contamination of the panels either side, poor appearance and possible overstressing when subject to movement (Figure 5.15c).

The sealant should be extruded steadily into the joint so that it is forced against the back and sides of the joint, taking care to avoid entrapping air and forming bubbles in the seal. Where joints are appreciably wider than the nozzle of the gun, beads of sealant should be applied into the corners of the joint and then further beads applied between the back of the joint and the corner beads of sealant, building a layer of sealant across the joint and then building another layer of sealant in front of the first until the joint is filled (Figure 5.16). Care must be taken to avoid entrapping air between the beads of sealant as this could lead to weakness in the section and consequential early joint failure.

In sealing very wide joints it may be necessary to fill the corners and back of the joint and allow the sealant to cure partially before applying the remainder in order to prevent slump.

At the end of the run, the extrusion pressure should be released to avoid afterflow or run-on of sealant from the gun. On air-operated guns the pressure is automatically released as the trigger is released but on hand-operated guns it is necessary to release the grip on the piston rod by means of the pressure release catch.

94 *Resealing of Buildings*

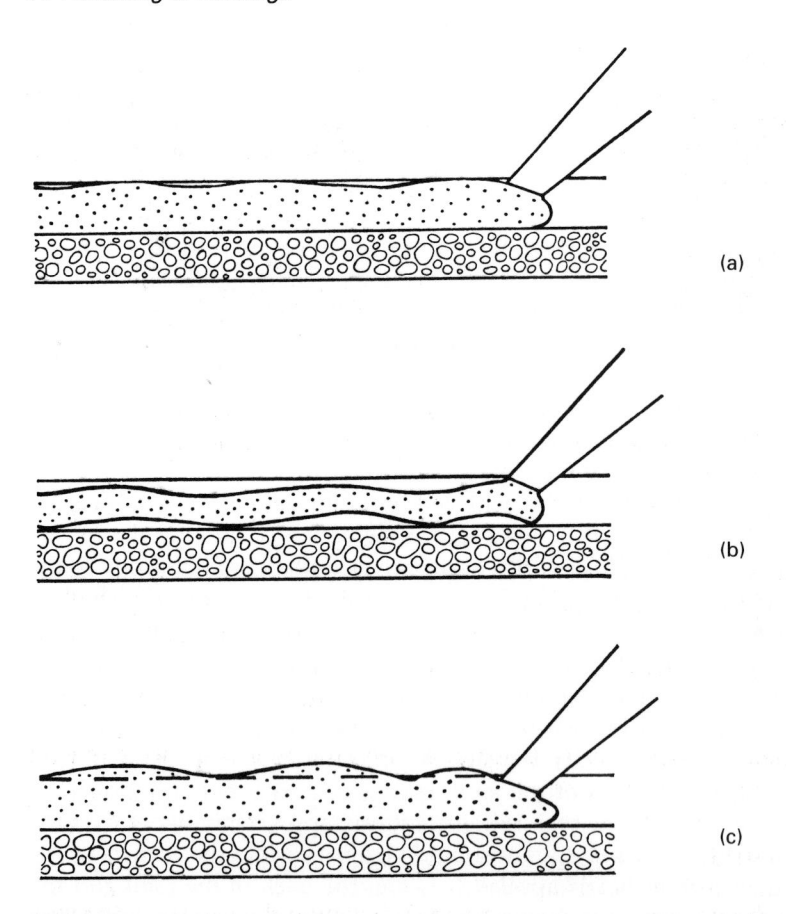

Figure 5.15 *Effect of gunning speed. (a) Correct speed-joint filled, slight undulations in surface; (b) too fast – joint underfilled, recessed with cavities; (c) too slow – joint overfilled*

A small amount of finely dispersed air occluded in the sealant during mixing has little effect on the sealant performance but can cause some afterflow of sealant from the gun. Larger bubbles trapped in the sealant pack will give rise to loud popping sounds when these bubbles emerge from the nozzle and burst. In the gun these bubbles are compressed to about a quarter of their original size by the extrusion pressure. This pressure is released as the bubble emerges from the nozzle and the consequential violent expansion causes the bubble to burst. This frequently gives rise to eruptions in the surface of the sealant but such disturbances are easily smoothed in the final tooling of the joint. Bubbles of this type are often trapped in the gun or pack

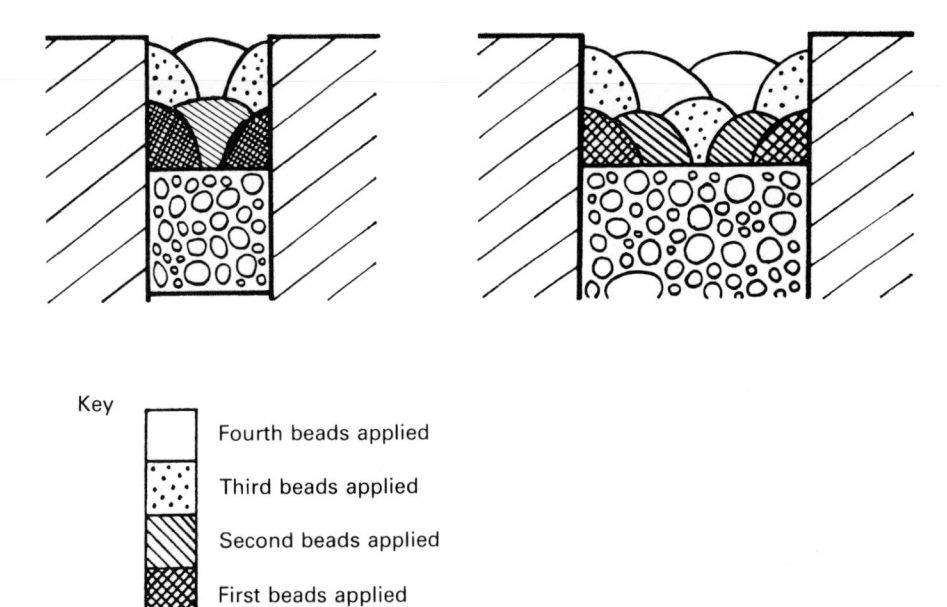

Figure 5.16 *Filling wide joints*

as it is filled, or they may be trapped as the nozzle or end-cap is refitted when changing packs or cartridges.

5.11 Tooling

Tooling is the process of smoothing and compacting the sealant in the joint while at the same time achieving a neat tidy finish to the joint as agreed or prescribed by the client. Tooling is carried out by drawing a suitably shaped piece of wood, metal or plastic over the surface of the sealant in the joint, pressing the sealant back to fill the back and corners of the sealant space, smoothing the surface and at the same time obtaining the desired surface profile of the seal (Figure 5.17).

Most joints are tooled to a smooth, slightly concave surface which is usually achieved with a smooth slightly rounded piece of wood kept wet with a dilute detergent solution. The choice of tool and wetting agent varies with the sealant and the applicator.

Tooling completes the sealing of the joint and helps to ensure that the sealant will perform correctly. The operation compacts the sealant in the joint, forcing the sealant into the back and sides of the joint and thus achieving the desired

Figure 5.17 *Tooling*

sealant section (*see* Figure 5.18). Forcing the sealant against the sides of the joint ensures wetting of those surfaces and also ensures maximum contact with those surfaces and hence maximum adhesion contact. Tooling the surface of the sealant smooths out the undulations due to the trigger action of the gun and also changes the bulbous convex surface to the flush or concave surface usually required. Various joint finishes are depicted in Figure 5.19. The action of tooling tends to eliminate bubbles and air pockets and by so doing, reduces the risk of weak points that could cause premature failure.

Tooling to a concave surface increases the ratio of contact area to cross section of seal and reduces stresses induced by movement. Tooling the seal to a concave surface improves the appearance of joints subject to movement. Sealants waist when stretched and bulge when compressed. Sealant joints finished flush with the face of the adjacent panels will show this effect usually to varying degrees, producing an unattractive appearance. Variation in the state of the joint at the time of sealing, slight variations in seal depth, degree of compression of back-up foam, and so on, can all cause differing degrees of bulge or waisting effects.

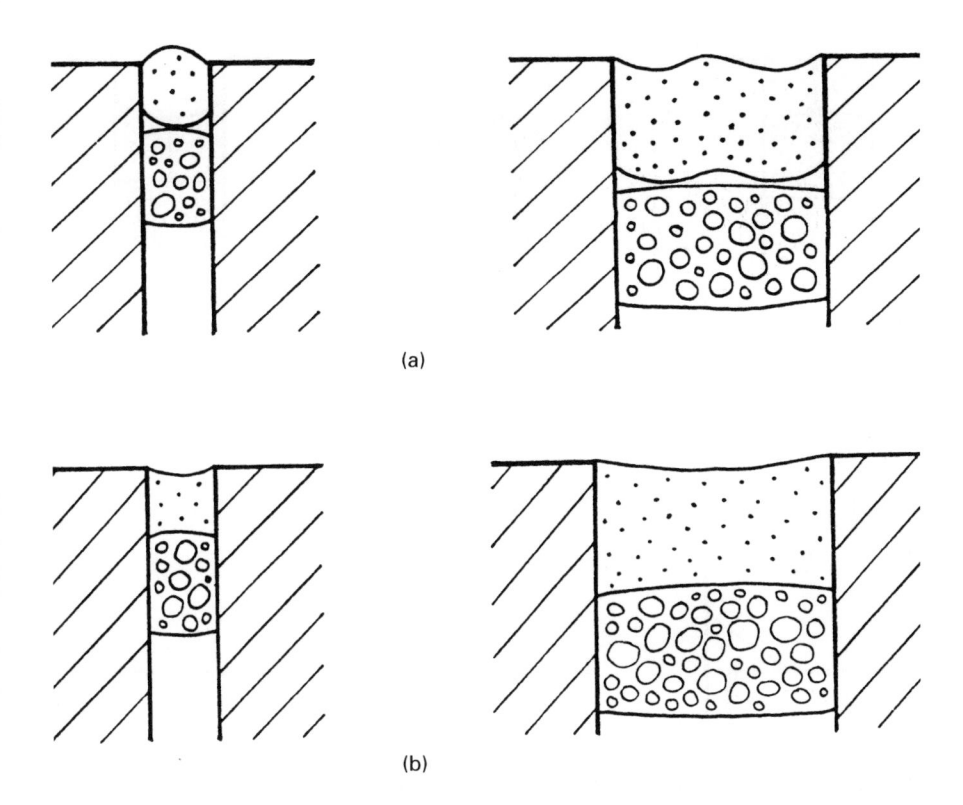

Figure 5.18 *Effect of tooling. (a) Before tooling; (b) after tooling*

Where flush-finished joints are subject to bulging, the surface of the seal may accumulate heavy soiling and may also suffer avoidable damage. Tooling the joint to a concave surface will improve the situation to a substantial degree but where joints are subject to large movements the joints should be recessed to minimize the visual effect.

Care must be taken to ensure an adequate depth of sealant at the centre of the joint because if it is tooled too thin it will buckle under compression and may be overstressed under tension. The seal at the centre of the joint should be in accord with the recommended width/depth ratio for the sealant used.

Adhesion of the wet sealant to the tool can usually be prevented by ensuring that the tool is kept wet with detergent solution or appropriate solvent. Softwood tools kept wet by soaking in either of these are widely used as they do not dry as quickly as metal or plastic, and they are easily formed to the shapes and sizes required.

98 *Resealing of Buildings*

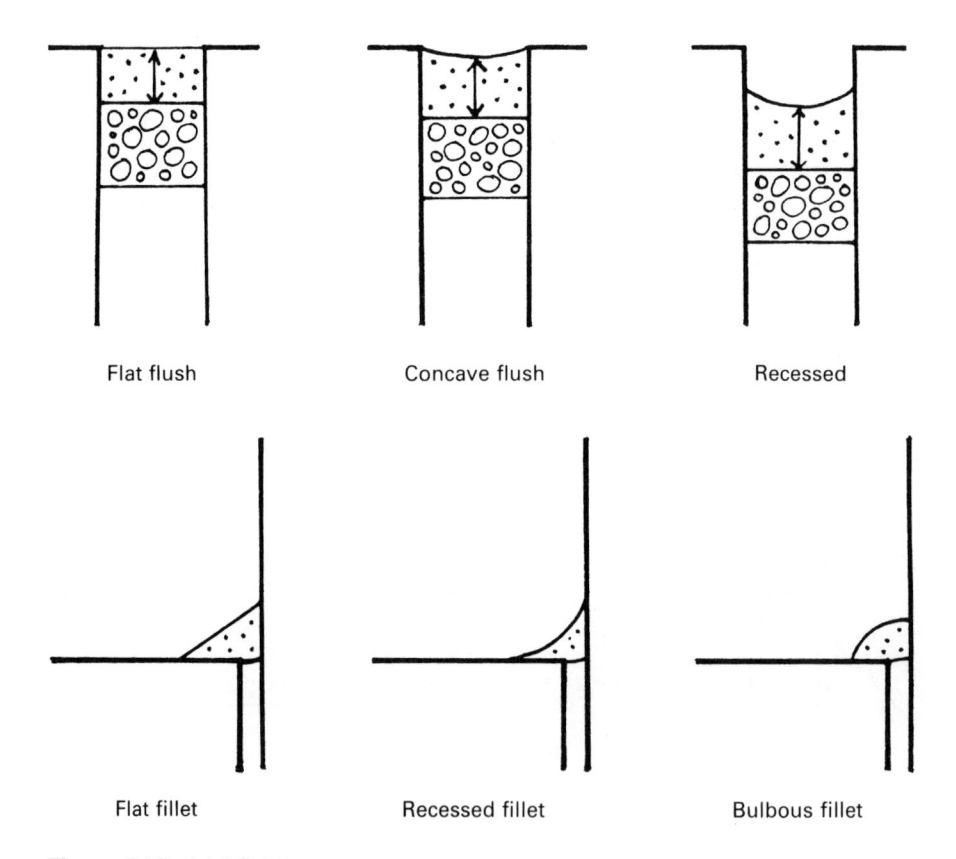

Flat flush Concave flush Recessed

Flat fillet Recessed fillet Bulbous fillet

Figure 5.19 *Joint finishes*

5.12 Masking

Where a flush joint or a specially neat appearance is required, the face/edge of the panels forming the joint may be protected by masking tape before priming, sealing and tooling the joint. Care should be taken to ensure that the masking tape is compatible with the surfaces concerned because some materials such as marble are easily marked. Joint priming requires special care as primers may penetrate under the edge of masking tape, if present, staining the surface and sticking the tape firmly to the sides of the joint.

Sealant should be applied to fill the joint space and then tooled to achieve the required finish. Various additional tooling techniques are used with masking, of which three are common:

1. Overfill the joint and use a broad blade scraper to tool the joint and remove the excess, scraping down the face of the joint leaving the joint with a flush, slightly concave surface.
2. Alternatively the scraper may be replaced with a rounded section of timber or plastic to give a deep concave finish.
3. Use a stiff flexible plastic scraper which gives a smooth finish when used in conjunction with a detergent spray. This method can also be used on smooth non-porous surfaces without masking tape as the polyurethane wipes the surface clean.

After tooling the masking tape should be removed before the sealant has cured, by pulling the tape across the sealed joint to ensure that any tails of sealant fall into the seal rather than onto the substrate either side (Figure 5.20).

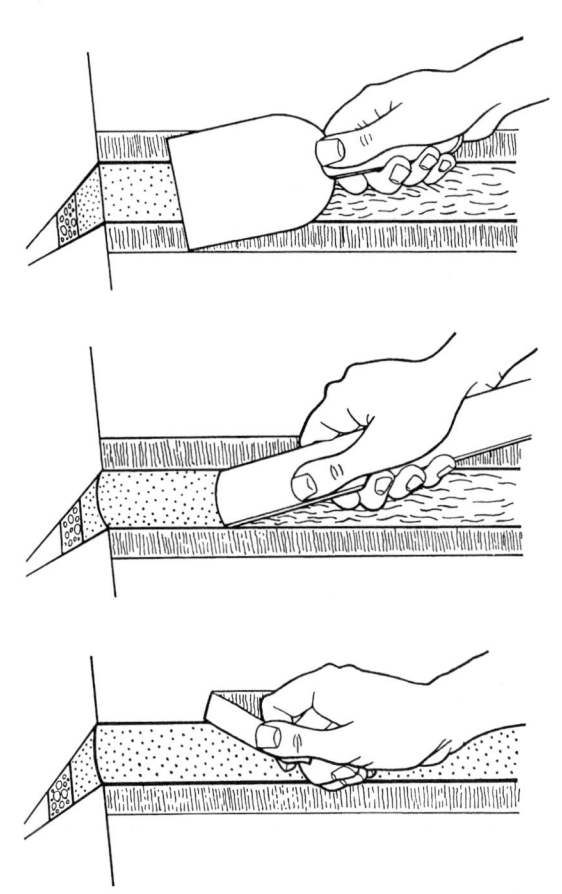

Figure 5.20 *Tooling with masking tape*

Any such tails should be gently tooled into the joint after the masking tape has been removed.

5.13 Protection during cure

Most joints in cladding do not require any special protection against weather during the cure period, but joints liable to physical damage such as floor joints may require protection during the early cure period. In many cases *covering the joint may retard the cure.*

Most two-part sealants will cure when covered but, as light and moisture often have a catalytic effect, full attainment of properties may be slower when covered. Solvent loss will be retarded by covering so that products that contain solvents or other volatiles will take longer to dry out and achieve their full properties.

Covering joints sealed with one-component materials is likely to have a more inhibitive effect. Heat and light have a catalytic effect in most one-part cure systems, but air and moisture are essential parts of the chemical cure process and their availability has a controlling effect on the rate of cure. Similarly, volatile by-products produced during cure have an inhibiting effect and if these are not removed by ventilation, they will slow or even stop the cure.

If joints sealed with one-part sealants need protection, any covering must be arranged to allow adequate ventilation to the joint and it may be necessary to allow for some delay in achieving full cure.

> *An example of this problem was a zoo aquarium occupying most of a building with a walkway around with viewing panels. The tank was lined with GRP with thick acrylic panels set in the side in a silicone sealant. The whole building was then closed up to ensure that the seals were not damaged or disturbed. When it was opened three weeks later, the building reeked of acetic acid and the sealant was in a soft sticky state showing only a very slight degree of cure. The building was left open with fans to circulate air around the viewing corridors and full cure was achieved in a few days. Poor ventilation, lack of moisture, and the inhibiting effect of the acetic acid produced as the sealant cured, had stopped the cure at an early stage.*

Where protection against mechanical damage and traffic is necessary, the joints should be protected with covers supported on blocks or packing to prevent contact with the sealant and to allow adequate ventilation to the sealant.

5.13.1 Chemical attack

Sealants have varying degrees of resistance to chemical attack. Cured sealants are generally resistant to solvents and other chemicals but the resistance develops as the cure progresses. In the uncured or part-cured state the sealant is vulnerable to attack. In such situations it is essential to prevent chemical attack until the cure has reached a resistant condition. In such situations it is often difficult to protect the joint itself and the only practical solution is to keep the unit out of service until an adequate cure has been achieved.

In a typical situation such as a swimming pool the sealant is susceptible to attack by water, oxidizing agents such as sodium hypochlorite, chlorine, bromine or ozone and other additives in the pool water. Few flexible sealants can withstand these in the uncured state but some polysulphides and polyurethanes are resistant when cured. Generally a fourteen day cure under warm conditions is necessary to achieve an adequate cure before the pool can be filled. Unfortunately this cure period is often not taken into account in the job programme, with consequential early failure of the joints when the pool has been filled too soon.

The cure process can also be affected by chemicals introduced to the environment for other purposes, such as the cleaning of floors with sterilizing agents, or the cleaning of brickwork with acid-based brick cleaners. These materials can interfere with the cure and in some cases prevent cure.

Where chemical resistance is required, it is essential to establish the minimum cure period and conditions necessary before the seal can be put into service and ensure that these are adequately allowed for in the resealing programme.

6 Quality Assurance guidelines

The Quality Assurance concept needs to be applied to the whole process of resealing to ensure that the intended design life is achieved. Every phase of the project, from initial inspection to final tooling and protection, needs to be considered and the product of each phase assessed to ensure that it makes its proper contribution to the overall performance of the resealed joint. The headings appropriate to Survey, Specification and Assessment considerations are contained in Appendix 1.

6.1 The survey

The survey must provide the basic information from which the joint design and potential reseal specifications can be derived. The survey report should include:

- A general description of the project and location, including joint sizes, width, depth, maximum and minimum, range and scatter of sizes and location.
- State of joints, damage, spalling, etc.
- Nature and age of existing sealant(s) and any other contamination.
- Cause of failure.
- Joint function(s).
- Joint substrates and component sizes.
- Design movement if known.
- Causes, types and amplitude and frequency of movements affecting the joint.
- Identification of any pattern of failure or any other non-standard situations. Where failure appears to be due to excessive movement, measure movement if possible.

6.2 Assessment of joint requirements

The first phase of joint design involves the assessment of the sealant performance requirements for the joints and the comparison of the requirements with the performance of the available sealants. In the case of early failure the original design requirements may still apply but, with later failures, new or modified conditions may apply due to structural or other permanent changes in the building or its component parts. This reassessment of sealant performance requirements may affect the option to use a similar sealant to the original or to use a different type.

6.3 Assessment of sealant suitability

Short-list sealants to accommodate frequency, type and amplitude of movement expected (see Sections 2.2, 2.3, 2.4). Having short-listed the available sealants suited to the joint performance, these must then be assessed in terms of their compatibility with site conditions.

- Will they adhere to the substrates with or without primers?
- Do they require different primers within one joint?
- Is the new sealant compatible with residues of the old that may be left in the joint? It should be remembered that performance will be reduced when adhering to old sealant residues (Section 5.2.2).
- Does the new sealant require special cleaning or preparation of the joint surfaces and are such requirements practical, bearing in mind any limitations of access, etc?
- Does the sealant require any special application procedure and is it compatible with site conditions? (e.g. Hot-applied sealants, high pressure application, etc.)
- Do the sealants have any limitations of weather conditions that may affect the resealing operations, that is maximum or minimum temperatures, susceptibility to rain, humidity or freezing?
- Are the sealants compatible with other materials or procedures on site at the same time? For example, volatiles from some materials may affect the cure of others or cause other effects, whilst acid cleaners prevent cure of one-part polysulphides.
- Are they appropriate to the type, amplitude and frequency of movement?

6.4 Site trials

Site trials are frequently essential to determine the suitability of a sealant and/or preparatory system. They are especially important if the joint surfaces may have been given a surface treatment or may be contaminated with oils or fluids.

Simple finger peel tests (*see* Section 4.3) are usually the most suitable with the requirement that adhesion should exceed cohesion after prolonged wetting or water immersion. Where possible, sample sections of the surface should be taken to the laboratory for full immersion tests but where this is not possible on-site tests including water immersion or continuous spray should be included (Section 4.3.1).

6.5 The reseal specification

As a result of the above selection process, it should be possible to produce a detailed specification for resealing the joints (e.g. see Figure 4.4). The specification should include:

- Minimum joint size, i.e. the minimum joint width necessary to accommodate the anticipated movement (Section 1), as measured or as calculated using the specified sealant (Section 2.2.1).
- Maximum joint size, i.e. the largest width of joint that can be sealed without slumping (manufacturer's data sheet) or the widest joint that is aesthetically acceptable. The maximum width may also be restricted by the depth of the joint and thus the limitations on depth of seal and thickness of foam backing that can be accommodated (Section 5.8).
- Sealant width/depth ratios (Section 5.8) to suit the anticipated range of joint sizes.
- Method of removal of existing sealant materials or other contamination and standard of cleaning required (Section 5.3).
- Method of final cleaning and priming if required (Section 5.3.3).
- Type of joint back-up material to be used, sizes and shapes to suit the expected range of joint sizes (Section 5.8.2).
- Sealant specification, material, section, width and depth, recessed or flush joint.
- Tooling specification (Section 5.11).
- Details of unsuitable weather conditions and any other limitations on application (manufacturer's advice).
- Permissible variations and procedures to deal with non-standard joints (*see* Section 4.1.3).

6.6 Operator technique

Within practical limits, the processes of cleaning and preparing the joints, together with preparation and application of the sealant, should be defined to ensure that the operator is fully aware of the process and the standard required. These should include:

- Method of removal of existing sealant materials or other contamination and standard of cleaning required.
- Method of final cleaning and priming if required.
- Type of joint back-up, size and shape to be used, position and method of insertion.
- Method of mixing sealant if required.
- Method of applying sealant and technique for larger joints where necessary.
- Section of sealant to be applied.
- Joint configurations: recessed or flush.
- Method of tooling and finish required.
- Type of joint protection, if required, and period for which protection is necessary.
- Chemical protection needs where appropriate.

6.7 On-site storage requirements

6.7.1 General

Cleaning solvents, primers and some sealants contain flammable materials and therefore require appropriate storage facilities on site (*see* Safety Data sheets). Some products come within the petroleum regulations, requiring special stores and limiting the amount that can be kept.

Some sealant materials are water-based and require protection against frost. Most materials will deteriorate more quickly if stored under hot or humid conditions.

Care must be taken to ensure that appropriate storage conditions are available on site to meet any regulatory requirements and to keep materials in good condition throughout the project.

Where materials having a limited shelf life are used, care must be taken to ensure proper stock rotation, and any life-expired material should be discarded.

6.7.2 Waste disposal

Storage and disposal of used packages and waste products should also be considered, especially where those materials may be hazardous or toxic. This may be especially critical on 'environmentally friendly' sites.

6.8 Health and Safety (COSHH Regulations)

Almost all chemical products are potentially hazardous. Sealant manufacturers strive to ensure that their products are safe and without risk to health when properly used for the purposes for which they are intended.

Most manufacturers give general health and safety information in the relevant technical data sheets and more detailed information is available in special health and safety data sheets available on request. Copies of the health and safety data on each of the products in use, including primers and cleaners, should be available on site at all times.

6.9 On-site quality control

6.9.1 Opening remarks

Joint sealing and resealing do not lend themselves to traditional methods of quality control as applied on the building site, by inspection on completion. Except in cases of blatant defects, faults are not obvious to superficial examination of the finished joint.

Cutting out sections of the joint may reveal faults that may be present, but it destroys the joint and the test area then requires resealing. Such destructive testing is undesirable as it introduces weakness and recreates the original problem.

As defects in joint preparatory work are covered but not corrected by subsequent operations, it is important to establish that the preparatory work is carried out properly at each stage. Detailed inspection at each stage would be costly and time consuming. It could also be counter-productive since joints may be left open for longer than is desirable awaiting inspection. This could result in contamination and other problems and could cause defects in adhesion which may induce early failure.

Retrospective or continuous on-site inspection is therefore not a practical method of quality control and testing on completion is destructive and undesirable. Quality control is therefore primarily a responsibility of the contractor carrying out the joint sealing operation. Quality depends on the operator doing his work well. This can usually be achieved by ensuring that the operator is aware of, and appreciates, the need for and the standard of work required, and that it is within his ability. This, together with adequate supervision, should ensure that quality is maintained.

It is important that the applicators and supervisors shall have an adequate knowledge of the principles of joint sealing and a detailed appreciation of the specification of the job in hand. The applicator should appreciate the importance of achieving the specification requirements at each stage of the work and that any shortfall could result in reduced quality and reliability of the installation. It is an advantage if the same applicators, of proven capability, complete the project.

Regular on-site supervision and inspection is essential, inspecting work at all stages to ensure that all aspects of the work are carried out to the required standard. When in doubt, testing by cutting out sections of finished joint, sealant and backing, should be included to ensure that there is an adequate seal depth, and checking seal adhesion. When sections are cut out, it is also advisable to check that the seal construction is in accordance with the specification. When making good, the seal must be reinstated to the required standard.

The principal areas in need of site assessment to ensure the quality of the seal are described below and contained in a check-list in Table 6.1.

6.9.2 Joint dimensions

The joint width and depth should be checked to ensure that they are within the limits laid down in the specification. This should ensure that provided the joints are sealed in accordance with the specification, a satisfactory installation should be achieved.

If the joint is too narrow, it may be possible to grind the sides to increase the joint width to the minimum required. If grinding is not practical, some alternative solution may be possible.

If joints are too wide it may be possible to amend the specification to use an alternative sealant or technique with the approval of the specifier.

Joint depth must be sufficient to accommodate the depth of sealant required as recommended for the width of joint for normal applications, or the increased depth required in load bearing situations, plus the thickness of back-up recommended for the joint width and situation.

If the joint is too shallow to accommodate the specified seal and back-up, it may be possible to reduce the depth required by substituting a rectangular section of foam for a rod form, or by substituting a bond-breaker tape for a rectangular foam provided that this is with the agreement of the specifier.

108 *Resealing of Buildings*

Table 6.1 *Check-list for site inspection*

1 Joint dimensions	Check:	(a) Minimum width (b) Maximum width (c) Joint depth
2 Joint cleaning	Check:	(a) Nature of original seal and/or contamination (b) Removal of original seal to achieve required standard of surface cleaning
3 Final preparation	Check:	(a) Cleanliness of surfaces in accordance with specification (b) Removal of dust or contamination by wiping with a clean cloth, solvent damped if necessary (c) Weather conditions are suitable for application of sealant system
4 Priming	Check:	(a) Primer used (as specified) (b) Primer applied in the correct areas (c) Primer drying time/open time (d) Reseal time
5 Back-up materials	Check:	(a) Type and size in accord with specification (b) Foam installed without damage (c) Position of back-up to ensure correct seal depth (d) Bond-breaker tape correctly installed if required
6 Application	Check:	(a) Correct sealant used (b) Joint filled to correct depth without entrapping air bubbles or other defects
7 Tooling	Check:	(a) Tooling to compact sealant, smooth surface removing blisters and irregularities, and to achieve the flat or concave surface required
8 Adhesion	Check:	(a) Adhesion to both sides of the joint using the finger peel test

6.9.3 Cleaning

Checks should be made to ensure that the cleaning procedure has been carried out and that it is achieving the standard required to obtain satisfactory adhesion of the new sealant.

In most cases visual examination and 'feel' of the cleaned surface is an adequate indication. Where silicones or greasy contaminants are present, a solvent wipe and adhesion tests may be necessary (*see* Section 4.3).

6.9.4 Final preparation and priming

It is first necessary to check that the weather conditions are suitable for application of the sealant system (*see* Section 4.2.3).

After major cleaning, but prior to sealing, surfaces should be wiped with a clean cloth or soft-bristled brush to remove dust, etc. If there is any possibility of greasy contaminants, the cloth should be moistened with a suitable cleaning solvent, and the joint face wiped towards the area of highest concentration.

After ensuring that the surface is clean, sound and dry in accordance with the specification, primers should be applied by brush or pad as recommended by the manufacturer and allowed to dry for the recommended time. Primers are usually visible either as a varnish-like coating causing a darkening water mark effect on porous surfaces, or as a dull slightly smoky transparent coating deposited by silane primers or surface conditioners.

Sealant must be applied after the primer has dried but within the recommended open time for that primer in order to ensure a satisfactory bond. It is not advisable to prime too far ahead in case delays may cause primer to be left too long to ensure good adhesion (Section 5.7 and Figure 5.3).

6.9.5 Back-up materials

Check that the back-up is of the type, quality, size and shape quoted in the specification and that it is correctly located in the joint without damage (*see* Sections 4.1 and 5.8).

Where bond-breaker tapes are used they should be suited to the sealant. Some sealants adhere to some tapes but not to others; for instance, most silicone sealants adhere to PVC tapes but not to polyethylene.

Bond-breaker tapes should adhere to the back of the joint, should cover the moving joint and should be at least as wide as the minimum joint width. The tape must not turn up the sides of the joint where it would reduce the bonding area of the sealant and could impose additional stress on the adhesive bond to the sides of the joint (Section 1.2 and Figure 6.1).

The back-up material should be set to leave the appropriate depth to accept the sealant section and where necessary to accommodate any recessing of the joint that is required.

110 Resealing of Buildings

6.9.6 Sealant application

Sealant depth is difficult to check without cutting out sections for examination and hence destroying the joint. It is, however, possible to assess the depth of uncured sealant by inserting a wire or needle into the uncured material and assessing the depth at which it encounters the firmer resistance of the back-up. The impression of the needle can be tooled out of the uncured seal.

When the sealant has cured with hollows and shallow areas, these can be detected by feel and the resistance to indentation. Hollow areas push in easily; shallow areas also indent more easily than deep ones but the difference is not so pronounced. A similar section of joint correctly sealed is

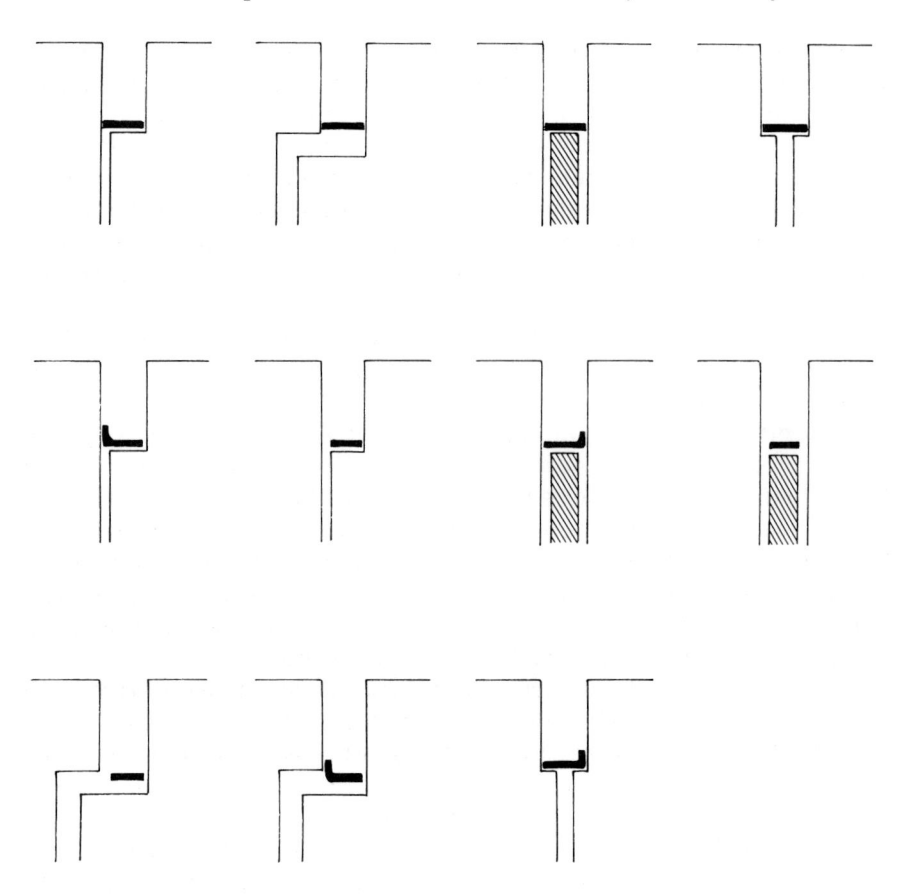

Figure 6.1 *Bond-breaker tapes*

useful for comparison purposes and the procedure benefits from the assessor having some experience.

It should be noted that some variation in depth is difficult to avoid, but substantial variations can result in uneven or excessive stresses in the joint causing early failure.

A shallow joint is likely to fold, crease, split and fail prematurely. A very deep joint may impose severe strain on adhesion and substrate, although a modest increase in joint depth is unlikely to cause any detrimental effect. If the superficial examination indicates that there may be a problem, a test section should be cut out to determine the actual depth of sealant.

6.9.7 Tooling

Tooling of the sealant is essential to achieve the maximum performance. Tooling compacts the sealant into the joint, filling the corners, eliminating air bubbles, improving wetting and adhesion, and smoothing the surface of the sealant to achieve the surface profile required.

Untooled or poorly tooled sealant may occlude bubbles and blisters, show poor wetting and adhesion and have a bulbous section leading to severe adhesion stress when the joint is subject to movement. The surface of the seal is likely to undulate with the varying extrusion pressure as the trigger of the gun is pulled.

A bulbous profile and an undulating surface of the seal are obvious indicators of the failure to tool the seal. Tooling must be completed before the surface of the sealant cures.

Two-part sealants cure throughout at a similar rate and tooling can usually be carried out throughout the application life and for some further time until the sealant becomes too rubbery and stiff.

One-part sealants cure from the surface inwards, first forming a skin and then the skin thickening and toughening as cure proceeds through the sealant. Tooling must be completed before the sealant forms a skin, otherwise the tooling action will crease, tear and rupture the skin producing an uneven, distorted surface that cannot be smoothed.

Tooling time varies from a few minutes with some silicone sealants to an hour or more with some one-part polysulphides. With short tooling time sealants, it is essential to apply sealant in short runs so that tooling can be completed before skinning occurs. With slower skinning sealants or two-part materials, a more lenient procedure can be adopted.

If the sealant has not been tooled, adhesion and compaction are suspect. Sections should be cut out and checked. If defects are found, the seal should be cut out and replaced.

If the sealant has been tooled but the surface is uneven it may be possible, depending on the material used, to overlay a thin layer of fresh sealant which can be tooled to achieve the desired finish.

If there is insufficient joint depth for overlay, it may be possible to cut back the surface of the sealant and then apply a thin layer of fresh sealant and tool to the required finish. In many cases it is better to remove the seal and reseal.

6.9.8 Adhesion

When the sealant has cured to form a rubbery mass, adhesion can be examined qualitatively by pressing the seal near the interface, pressing into the joint and away from the substrate. This imposes a peel stress onto the bond between sealant and substrate and gives a good indication as to whether the bond is sound. Care must be taken to assess the bond without damaging the joint. The adhesion strength should be greater than the cohesive strength so that it should be possible to press the seal until the sealant begins to tear, without damaging the sealant bond.

7 Case histories

The following case histories (summarized in Table 7.1) illustrate particular cases of joint seal failure, identification of the causes of joint failure and the remedial treatment carried out. It should be noted that this information is based upon the state of knowledge at the time of writing. Specific details of the buildings or products concerned are not identified.

Table 7.1 *Summary of case history material*

Case history	Subject
7.1	Large frequent cyclic movements in thin concrete cladding
7.2	Large cyclic movements in aluminium cladding
7.3	Staining of marble
7.4	Adhesion to oil-contaminated brickwork
7.5	Oversealing gasket glazing in curtain walling
7.6	Blistering acrylic sealant on hardwood
7.7	Oil-based mastic sealant discoloration from contact with damp-proof membrane
7.8	Replacement of hidden seals in curtain walling
7.9	Poor preparation of joints in precast concrete cladding
7.10	Contamination failure following resealing
7.11	Silicone sealant discoloration from contact with pitch-polymer damp-proof membrane
7.12	Extrusion of bituminous material from concrete joints in a swimming pool
7.13	Silicone sealant discoloration from contact with rubber thermal break in window frames
7.14	Sealant blistering caused by water-repellent concrete surface treatment
7.15	Chemical attack of sealant by brick cleaning fluid
7.16	Use of abseil technique
7.17	Smelly sealant
7.18	Staining by silicone sealants

7.1 Large frequent cyclic movements in thin concrete cladding

The problem

Many school buildings were erected using a building system incorporating concrete cladding panels and aluminium windows fixed to a structural steel frame. The concrete panels were thin lightweight panels approximately 1000 mm wide with insulation and plasterboard lining. A cold-applied acrylic sealant was used to seal the joints between panels, and between panels and windows. Joints were approximately 8 mm wide.

After several years some leaks were reported. Examination of the sealed joints revealed vertical hair-line cracks in many of the joints between the concrete panels. The acrylic sealant was still soft, pliable and easily deformed.

With care it was possible to cut out joint sections using a sharp wet knife. This showed that in many cases the surface crack passed through the sealant to the back of the joint and, although the crack usually appeared in the middle of the seal or to one side on the external joint face, it coincided with the edge of the backing foam at the back of the joint. Where two pieces of foam had been used, the crack coincided with the junction. In some cases the crack did not pass completely through the seal. Some joints exhibited no defects.

It was noted that in all cases the acrylic sealant was soft and pliable, adhering well to the sides of the joint and also adhering to the back-up foam. In most cases the foam was slack in the joint.

In the course of the investigation many similar buildings were found which had no leaks. Some were found with defects but no leaks had been reported. Two buildings had been sealed with alternative sealants, one with a hot-applied acrylic and one with one-part polysulphide sealant; neither showed any leaks. The hot-applied acrylic sealant was no longer available.

Consideration of cause of failure

In the course of this investigation it was noted that the concrete panels often felt warm where they were exposed to the sun – much warmer than normal concrete panels. Investigation revealed that the panels were about 75 mm thick at the edges but sustantially thinner over the majority of their area. In these buildings there is a cavity behind the panels and more than 100 mm of fibreglass insulation and a plasterboard lining. The concrete panels have a much lower heat capacity than normal thick, heavy concrete and will therefore heat up quickly. Since they cannot dissipate that heat into the building because of the insulation, they will get warmer than normal.

Generally speaking, concrete panels would normally induce small, very slow cycles of movement. However, in these buildings the very thin panels, combined with the construction and insulation of the building, gave rise to both high frequency cycles of movement and larger movements than would normally be associated with lightweight cladding systems.

With such movement patterns, a more elastic sealant is desirable. The more elastic hot-applied acrylic is better in such situations than the cold-applied sealant; a one-part polysulphide sealant is, however, better still.

Remedial treatment

Repairs were carried out in two ways. Some buildings were repaired by cutting out the existing seal and back-up, replacing the backing with a larger polyethylene foam section, compressed into the joint and set to allow a deeper seal about 8 mm wide by 10 mm deep. This technique was considered as complying with the original specification and avoided the need to remove all traces of the old sealant. It has been successful for over 8 years.

Other buildings were repaired by cutting out the existing seal and scraping the residues from the joints as far as possible. The joint faces were then cleaned with a toluene-based cleaner and primed for the new sealant. New over-size polyethylene foam backing was compressed and inserted to give a seal depth of at least 10 mm, and the joints resealed with one-part polysulphide sealant and tooled to a slightly concave surface. The polysulphide sealant has performed satisfactorily for more than 12 years.

7.2 Large cyclic movements in aluminium cladding

The problem

This building was essentially a concrete-framed office block clad with aluminium windows and pressed aluminium panels, finished with a matt charcoal grey paint. The building was designed to be energy-efficient, having a very high standard of thermal insulation throughout.

Problems arose with the joints between the aluminium cladding panels. The panels were about 1800 mm wide and 1000 mm high, with 15 mm vertical joints and 10 mm horizontal joints. All joints were sealed with a two-part polysulphide sealant over a polyethylene foam backing.

Problems appeared soon after the cladding had been fixed and sealed. Bulges appeared in the joints, giving the impression of poor uneven application and the sealant applicator had to replace many metres of joint before the work was accepted by the clerk of works. Despite this, about a year after completion, the joints were reported as bulging and uneven and some areas were replaced again.

Several years later, when the building was about to be sold, there were reports that the joints were deformed, split and leaking. Examination revealed that sealant in the vertical joints was bulging forward of the panels and tending to split to one side although adhesion to the panels was excellent. Sealant in the horizontal joints was also bulging but to a lesser degree.

Examination of sealant samples confirmed that it had been mixed satisfactorily and that the original seal depth had been adequate. Similar cladding had been used on other buildings elsewhere, sealed with the same sealant and was performing satisfactorily.

The cause

It was only when the cladding details were linked with the construction details that the unusual features were apparent. Behind the cladding was an exceptional amount of insulation; 300 mm of lightweight insulating block and 100 mm of mineral wool was common.

Even on cold days, the building produces a heat haze in front of the cladding. Temperatures in excess of 80°C have been measured on the face of the cladding on sunny days. On sunny days it is possible to stand in the car park, 100 metres away, and listen to the shuffling noise as the cladding moves when clouds pass in front of the sun.

The cladding on this building is thermally isolated from the building itself. Heat absorbed by the cladding cannot be dissipated into the building but is retained in the cladding. The matt charcoal-grey finish is an excellent heat absorber and aluminium has a low heat capacity. The temperature of the aluminium rises quickly when the sun shines on the cladding. As a result the cladding is subject to rapid and frequent changes in temperature, and to extremes of temperature. These changes in temperature produce expansion and contraction of the aluminium cladding components and consequential opening and closing of the joints.

The joints were sealed with a two-part polysulphide which was elasto-plastic in nature and therefore not well suited to accommodate this type of movement. Such sealants deform and split when subject to large, frequent cycles of movement. A more elastic sealant is necessary to accommodate this type of movement.

Remedial treatment

The existing sealant was cut out and the joint faces cleaned to remove all traces of residue. The aluminium finish was a stoved acrylic and was not damaged by toluene cleaning solvent. New polyethylene foam was inserted and the joints resealed with a low modulus silicone sealant.

Special problems occurred in the resealing as the face of the building became intolerably hot and it was only possible to work in the early morning. Most work was done between 5 a.m. and 10 a.m. as the face of the cladding became too hot thereafter.

The new sealant has performed satisfactorily for more than 15 years.

7.3 Staining of marble

The project concerned was a shopping complex, part of which was clad in a novel type of marble cladding. The marble was in the form of 6 mm thick sheets bonded to an expanded aluminium web, which was in turn bonded to an insulating lining board. These composite panels formed a lightweight cladding giving the appearance of marble but at much reduced cost and weight.

The cladding was to be weatherproofed by sealing the butt joints between the marble panels. The cladding was sealed with a polyurethane sealant, applied without primer to avoid the risk of primer marking the stone.

After a few months a grey shadow effect was observed either side of the joints, caused by plasticizer migration into the stone from the sealant. The shadow progressed from the edge of the joint, extending into the stone and becoming more diffuse (Figures 7.1 and 7.2).

The cause of the problem

The cause of the problem was the combination of this particular polyurethane used on unprimed marble. Plasticizer migration from polymeric sealants is difficult to predict and even more difficult to reproduce. It does not occur on all porous stones; plasticizer migration is seldom seen on limestone or reconstructed stone. Marble is however particularly susceptible to plasticizer migration. Plasticizer staining also occurs occasionally on granite which has a low porosity. Slate will stain from some edges more than from others. The staining appears to be an interaction between stone and plasticizer in which plasticizer is drawn into the stone from the sealant by some form of capillary attraction. Temperature and pore-size play a part and there is some indication

118 Resealing of Buildings

Figure 7.1 Staining of marble (case history 7.3)

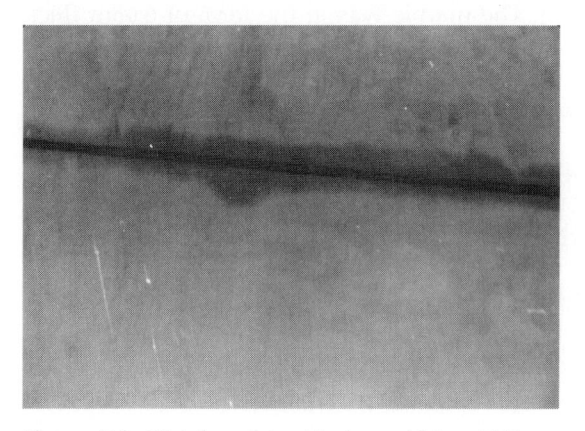

Figure 7.2 Staining of marble (case history 7.3)

that joint movement, particularly compression, may also influence the migration.

Marble is very susceptible to migration staining. Complete sealing of the edge is probably the only way to prevent staining. However, in the project concerned this may be impossible. The nominal joint width is 6 mm and the marble layer is 6 mm thick. The seal must be at least 6 mm deep so that the back-up foam must be supported by the panel edge behind the marble. The aluminium grid bonded to the back of the marble has holes more than 25 mm across. At the edge of the panel the grid is intermittent so that when the back-up foam is inserted it is only supported where it bears against the grid and is loose in the gaps between. As a result, most of the back-up foam is poorly supported and it does not offer a firm support to the sealant. As sealant is gunned into the joint the foam is likely to be displaced, allowing sealant to pass around the back of the marble layer onto the back face of the stone panel. Although it may be possible to seal the edges of the marble by careful application of primer through the open joint, there is no way of sealing the back face of the marble with the cladding in place. If any sealant is in contact with unsealed marble, migration staining can occur.

Removal of the existing stain is another problem. The stain is inside the stone and is probably at least as deep as the joint. It may have already traversed many millimetres across the panel. Although there is little chance of removing the migrant plasticizer, it may be possible to reduce the intensity of the stain. If the source of the stain, that is the sealant, is removed, the intensity of the migrant stain may be reduced by heat or by thorough washing with an appropriate solvent.

Remedial treatment

There is no easy or satisfactory solution to this problem.

The basic difficulty is with the original design. The stone face is too thin to allow an adequate joint incorporating primer/sealer to prevent migration staining, or to accommodate an adequate joint backer, properly held in place, and to allow an adequate seal depth with the seal recessed to conceal the primer/sealer.

With the existing detail everything is inadequate and the result will always be in doubt. No remedial treatment could be guaranteed.

7.4 Adhesion to oil-contaminated brickwork

This project concerned the refurbishing of a 15-year old housing development. The anodized aluminium windows were set into brickwork surrounds and perimeter-sealed with a sealant that appeared to be a polymer-modified oil-based mastic. The same mastic had also been used to seal structural joints in the brick cladding.

There were gaps about 2 mm wide between mastic and substrate around the windows and larger gaps in the brickwork joints. Perimeter clearance around windows was about 8 mm whereas the brickwork joints were 12–15 mm wide.

The problem

The mastic had become hard, dry and leathery and had lost adhesion to both surfaces to a considerable extent. The mastic could be separated from both the brickwork and the metal, leaving the surfaces relatively clean but the brickwork remained discoloured and stained by oil leached from the mastic. The oil stain was not removed by solvent cleaning.

Consideration and testing

The client wanted a sealant with a life expectancy of at least 20 years. The sealant contractor tried various methods of cleaning and found that cutting and scraping with knives and wood chisels removed most of the old sealant. Brushing with a stiff brush and wiping with a solvent cloth substantially removed the remainder but left the oily stain on the bricks.

Polysulphide and silicone sealants were tried on site, with and without primers, and tested by the finger peel test after curing for 3 weeks and water spray for 24 hours. Most of the sealants gave good adhesion to the anodized aluminium without primer. Adhesion to unprimed brickwork was generally poor, especially after water spray. Adhesion to primed brickwork was better and was generally maintained after water spray.

Remedial treatment

All joints were prepared by cutting out the original mastic and scraping with knives and wood chisels to remove the bulk of the hardened mastic. This was followed by brushing with a stiff brush and wiping with a clean cloth moistened with a toluene-based cleaner. The cloths were changed regularly.

The brickwork was primed with a resin-based primer and allowed to dry. New polyethylene foam backing was inserted to leave the correct joint depth and the joints sealed with a low modulus silicone sealant specially selected for its ability to stick to oily surfaces.

The project was completed in the mid 1980s and was completely sound at the time of writing.

7.5 Oversealing gasket glazing in curtain walling

The problem

The project was a 10-storey office block near to the coast and facing down the Bristol Channel towards the Atlantic Ocean. The facade consisted of a bronze-anodized aluminium curtain wall with glass infill. All panels were glazed with gaskets using a drained and vented glazing system. About 12 years after completion there were reports of severe water entry in driving rain conditions; water entered at the bottom glazing lines, being blown through the system and landing some distance beyond the windows. In less severe storms water flowed over the bottom rebate and there was seepage at some transom/mullion joints.

Detailed inspection showed that the glazed panels were fixed with insert gasket sections either side of the glass. Removing a panel revealed that the cavity around the glass was drained through holes in the frame so that water entering the glazing cavity should drain to the outside. However, the glazing channel was partially blocked with sand, insects and other debris. The glazing gaskets appeared to have shrunk and there were gaps up to 50mm long at the corner joints both inside and outside the glass.

Vertical expansion joints in the facade appeared sound and there was no evidence of water entry associated with these joints.

Discussion

The shrunken gaskets allowed excessive water penetration into the glazing channels. The internal gaps allowed a through-flow of air in windy conditions, carrying water into the glazing through gaps in the external gaskets and blowing water through the gaps in the inside gaskets in strong winds. The silting up of the glazing channels prevented free drainage causing large quantities of water to collect in the channels, overflowing into the building and seeping through weak joints.

Remedial considerations

Clearing the glazing channels and replacing the gaskets would have restored the original conditions and prevented the ingress of water. However, this would have involved de-glazing, which would have drastically interfered with the use of the building and there was a probability that the same problem could occur again.

Oversealing the existing gaskets would prevent the ingress of water or sand and should prevent water entry. The drainage system, even if not very efficient, should be able to deal with any minor leakage. Overcoating the internal gaskets would eliminate the risk of a through draught and further reduce the risk of water entry.

The external sealing could be carried out from the existing window cleaning cradles or larger platforms without interfering with the occupants and the internal sealing could be carried out at weekends. This would minimize any inconvenience to the tenants.

Remedial treatment

The glazing gaskets were oversealed with a black low-modulus silicone sealant, internally and externally, thus maintaining the appearance of the original gaskets and avoiding any risk of discoloration of the sealant due to contact with the rubber gaskets.

Resealing was completed in the early 1980s and there have been no reports of water entry since that time.

7.6 Blistering acrylic sealant on hardwood

The problem

In refurbishing a residential estate, new hardwood (Filipino mahogany) window frames were installed in brick surrounds and their perimeters sealed with a solvented acrylic sealant. Within a short time the contractor was recalled to replace sealant where it had bubbled and blistered along the length of some of the joints. A few weeks later the contractor was again recalled because the new seals had blistered again, together with more of the original seals.

A detailed investigation of the site revealed that blistering occurred on the south and west elevations where the windows were exposed to the sun. Examination showed that the bubbles emanated from the surface of the timber and were confined to the timber side of the joint (Figures 7.3 and 7.4).

Site trials were carried out using a variety of sealants and surface pretreatments. These confirmed that the blistering effect occurred with a variety of sealants. Non-curing sealants were more susceptible than curing sealants, and the faster-curing sealants were the least affected.

Primers were effective in reducing the blistering but they did not eliminate the problem, because of the difficulty in applying a continuous film without pinholes or perforations.

The cause

The problem is caused by the cellular nature of the timber which allows air, moisture and solvent vapour, that is absorbed in the timber, to be forced out when the timber is warmed by the sun. This causes pressure which blows blisters in the soft plastic seal.

Figure 7.3 *Blistering acrylic sealant on hardwood (case history 7.6)*

Figure 7.4 *Blistering acrylic sealant on hardwood (case history 7.6)*

Remedial treatment

Remedial work involved the removal of all traces of the acrylic sealant by cutting out, scraping, solvent cleaning, and resealing with a fast-curing silicone sealant. The remedial sealant was applied late in the day to minimize the effect of rising temperatures.

7.7 Oil-based mastic sealant discoloration from contact with damp-proof membrane

The problem

Powder-coated aluminium windows were used in a new school extension and the perimeter joints sealed with a polymer-modified oil-based mastic. After about 12 months the contractor was recalled as large areas of the mastic had turned yellow.

Various suggested causes were investigated, including local environmental factors such as contamination from adjacent pine trees, but not proven. The bedding mastic used in the windows could produce discoloration but not to match that on site.

During the school holidays it was possible to cut out the defective areas when it was found that a black plastic dpc used around the window openings was in contact with the seal. The discoloration was due to bituminous materials from the dpc migrating into the sealant.

Remedial treatment

The seal was completely removed, the dpc was cut back so that it was completely isolated by the polyethylene foam back-up, preventing any contact between seal and dpc and the windows resealed with the same product.

7.8 Replacement of hidden seals in curtain walling

The problem

In the late 1960s it was firmly believed that protecting sealants against the weather would extend their life. This was a feature of the design for the curtain wall cladding for a city tower block, but experience has shown the difficulty of resealing the project.

The curtain wall was designed as factory-assembled panels fitted together on site with a double tongue and groove connection as shown in Figure 7.5. The inner tongue served to locate and align the panels. The outer tongue and groove formed the sealed joint, the groove was partly filled with sealant and the tongue of the next panel inserted into the sealant. Finally a top hat-shaped cover strip was clipped over the joint to hide the variable gap between panels and completely protected the seal from the weather.

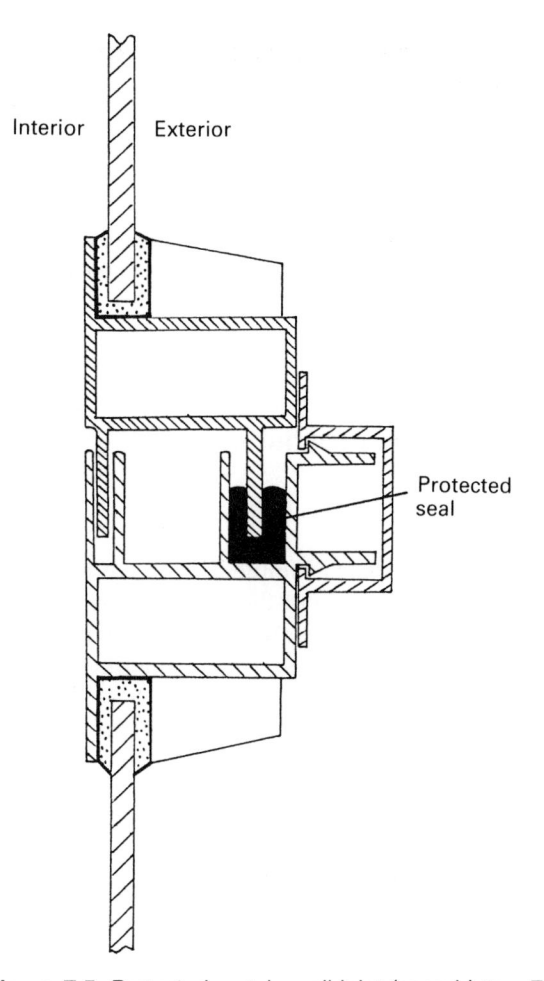

Figure 7.5 *Protected curtain wall joint (case history 7.8)*

The seal used was a two-part polysulphide but it was bonded on three sides, inhibiting its ability to accommodate movement. After about 12 years some water penetration was reported. The original seal cannot be replaced without completely dismantling the curtain wall from the building. Several attempts have been made to apply fillet seals between the cover strip and the panels but with little success. There is only an edge of metal to bond to, and such a seal cannot accommodate the movement of the cladding. The cover strip is held by clipping over nibs in the extruded aluminium sections and cannot be removed without causing damage to the cladding.

Remedial treatment

There is no means of repair without recladding, unless it is possible to remove the cover strips and fit new, redesigned, cover strips to accommodate new joints between cover strips and panels. In either case resealing will be very expensive and difficult.

7.9 Poor preparation of joints in precast concrete cladding

The problem

This case is recorded because it was a classic example of 'you get what you pay for'. Even in new work, a large proportion of the cost is in preparation. Low prices may mean undesirable economies in preparation which result in premature failure. The project concerned a large college building clad in precast concrete panels and anodized aluminium windows.

The sealing to windows was part of the window contract and gave no problems. The sealing of joints in the panels was let to the lowest tenderer. About three years later it showed many defects, including almost total adhesion loss where the concrete had not been cleaned or primed. Sealant was splitting where it had been applied too thinly: 3 mm thick in 30 mm joints. Splitting failures were observed where it had been applied directly onto polystyrene or fibreboard joint fillers without polyethylene foam back-up or bond-breaker tape.

The original sealant contractor could not be traced.

Remedial treatment

The remedial specification simply involved removal of the existing seals and replacing them with the same type of sealant, correctly applied. Joint fillers were cut back to allow insertion of the correct back-up foam, concrete surfaces were cleaned and primed in accordance with manufacturer's instructions and sealant was applied in the recommended width and depth.

The remedial work was completely successful and had remained sound for more than 10 years at the time of writing.

7.10 Contamination failure following resealing

The problem

The project concerned a four-storey office block, built around 1968, and clad with heavy precast concrete panels with exposed aggregate faces but smooth concrete arrises. The joints between panels had originally been sealed with a butyl-based mastic applied 20 mm by 10 mm onto polyethylene foam backing strip. After some fifteen years the building had developed numerous leaks in the cladding, affecting all four elevations but worst on the south side.

The initial reseal specification required: cleaning out of old sealant; wire brushing; priming; and resealing with a one-part polysulphide sealant. Work was completed but substantial failure occurred due to incompatibility between new sealant and residues of old butyl sealant. Joints 10 mm wide are difficult to clean adequately, especially on porous concrete.

The second reseal specification required: removal of all polysulphide sealant (not difficult, due to poor adhesion); cleaning out remaining butyl sealant as far as possible; removing all loose or friable material; and resealing with solvent-based acrylic sealant.

The second reseal has remained satisfactory for several years.

7.11 Silicone sealant discoloration due to contact with pitch-polymer damp proof membrane

The project concerned a five-storey office block fitted with new uPVC windows into brickwork openings. Window perimeters were sealed with a white silicone sealant. After a few weeks, sections of the sealant began to exhibit a yellow/brown discoloration which became progressively darker (Figures 7.6 and 7.7).

The cause

The pitch-polymer vertical dpc around the brickwork openings had been brought through too far and was in direct contact with the back of the silicone sealant. Constituents from the dpc had migrated into the silicone sealant to produce the brown stain at the surface.

Remedial treatment

The remedial work involved; cutting out all sealant from affected joints; cutting back the dpc; inserting a bond-breaker tape/backing strip to isolate

128 *Resealing of Buildings*

Figure 7.6 *Silicone sealant discoloration from contact with pitch-polymer damp proof membrane (case history 7.11)*

Figure 7.7 *Silicone sealant discoloration from contact with pitch-polymer damp proof membrane (case history 7.11)*

the dpc from the sealant; and resealing with a white low modulus silicone sealant.

7.12 Extrusion of bituminous material from concrete joints in a swimming pool

The project concerned was a swimming pool, and the problem was described by the manager as 'black sticky stuff' coming up through the joints in the pool. This was making a sticky mess in the pool, contaminating bathers and swimming costumes, resulting in many irate customers, claims for damage and consequential closure of the pool.

Site investigation confirmed the manager's story. The sealant used to seal the expansion joints in the tiling was a chemically-resistant two-part polysulphide and the 'black sticky stuff' appeared to be squeezing up between the sealant and the tiles. Analysis of the 'black stuff' showed that it was bituminous and therefore could not originate from the sealant. It appeared to come from somewhere below the seal in the tiling.

Investigation revealed that the joints in the shell of the pool had been formed with a bitumen-bonded cork board and sealed at the surface with a hot-poured bituminous sealant. The tiles were laid onto a thin-bed adhesive and the joints located over the joints in the concrete shell and separated by a thin bond-breaker tape.

When the pool was heated, expansion of the concrete caused closure of the expansion joint, compressing the filler board and the bituminous sealant. As there were no voids in the joint to allow this deformation, the sealant was squeezed under the tiles and extruded through any gaps, holes or weakness that it could find. Detailed inspection showed that the 'black stuff' was extruding through gaps and weaknesses against the tile joint but it was also appearing through small gaps in the tile grout.

Remedial treatment

The pool had to be drained and the tiles removed either side of the joints so that the joint in the concrete pool shell could be recessed. This allowed the insertion of a square section of polyethylene foam beneath the tile joint to allow the joints to deform without creating the high pressures that caused the extrusion of the bituminous material. The cleaning and drying took a considerable time but there have been no reports of any further problems.

7.13 Silicone sealant discoloration from contact with rubber thermal break in window frames

A new office block in the north-west of England was fitted with white thermal barrier aluminium window frames. The perimeter joint was sealed with a low modulus silicone sealant. A few months later there were reports of the sealant going brown. It was confirmed that there was no bituminous or pitch-polymer dpc in the surround. It was also found that when the discoloured surface was cut away from the seal, it was white underneath but the newly revealed surface soon discoloured like the original.

When sections of sealant were cut out on site it was found that there was a thin sliver of black rubber cut away from the edge of the frame. The frame was formed from an inner and an outer section screwed together with a synthetic rubber isolating strip between the sections to form the thermal barrier.

The rubber isolating strip was the cause of the problem. Where the silicone sealant was in contact with the rubber, discoloration appeared on the surface. Laboratory tests confirmed the effect, which appeared to be due to antioxidants or curing agents used in the rubber which migrated into the silicone and discoloured when exposed to light. The rubber was black and would not exhibit any discoloration.

Remedial treatment

The silicone sealant was cut out, taking care to remove all traces. An oversize polyethylene foam backing strip was inserted and carefully positioned so that it prevented any possibility of the new sealant coming into contact with the rubber isolator. The joints were then resealed with the same white low modulus silicone sealant.

No further discoloration has occurred.

7.14 Sealant blistering caused by water-repellent concrete surface treatment

The problem

The project concerned refurbishment of an office block in the south-west of England. The original cladding consisted of concrete mullions and panels with windows over the panels between mullions. The panels had deteriorated and were being covered with 'Granitex' panels, with the new panels sealed to the concrete mullions using a one-component polyurethane sealant.

The work was carried out in good conditions but widely varying temperatures, cool overnight but warming up rapidly in the autumn sunshine. Complaints of blistering were received within a few days. Some of the early work was cut out and replaced with new sealant, but it soon blistered again.

Samples of sealant were gunned into aluminium channels and exposed alongside the defective joints. No blisters appeared in the channels, confirming that there was no fault in the sealant. It was concluded that the blistering was due to some interaction between sealant and joint.

Sections of sealant were carefully cut out of the defective joints. These showed that in all cases the bubbles emanated from the sealant/concrete mullion interface. There were no bubbles on the 'Granitex' panel side of the seal, whereas large blisters had formed on the concrete mullion side. Site tests carried out with an alternative sealant, a one-part polysulphide, produced joints that did not blister.

Further investigation revealed that the concrete surface had been treated with a clear water-repellent finish some years previously although the precise identity of the coating was not known. When they deteriorate, some clear water-repellent finishes are believed to cause moisture retention. The external surface becomes less repellent, allowing moisture to penetrate, but the coating that has been absorbed into the body of the stone is protected from the weather and does not deteriorate. Water that has been absorbed by the concrete is repelled from the surface and remains trapped within the body of the stone or concrete.

Examination of the mullions showed that the surface in the jointing area was pock-marked, blow-holed and uneven. The blisters in the seal appeared to emanate from these blowholes and irregularities in the mullion surface. Moisture was present in the bubbles, and the surface reinstatement to seal the blowholes was considered to be uneconomic.

Consideration of cause of failure

When the mullion surface is exposed to heat from the sun, moisture trapped within the body of the concrete will expand and vaporize. If it cannot escape through the surface, it will find some other exit such as blowholes, cracks and untreated areas at the edge or areas where the coating may have been removed.

Moisture escaping in this way can develop considerable pressure and may blow blisters by pressure alone; alternatively the moisture can react with some sealants. Some polyurethane sealants may be prone to blistering as moisture can react to produce carbon dioxide gas.

In this situation there are several factors that could contribute to the blistering that occurred. The primary factor was the concrete coating which created conditions that could cause blisters to develop in the seal.

The project was resealed with an alternative sealant that did not blister. The primary lesson is the need to carry out site trials under site conditions, especially where coatings have been applied to porous materials such as stone or concrete.

7.15 Chemical attack of sealant by brick cleaning fluid

The problem

This case involved a prestigious office block in the south-west of England completed in 1986. The following spring it was reported that there was a strange smell in one of the inner courtyards. The smell was strongest in hot weather and was initially attributed to fumes from the ventilation system. However, the heating engineers could not find any fault with the system. Over the next two years, and further investigations of the heating and ventilation system and the drains, the source of the smell was identified as the seals around the windows. The sealant manufacturer and contractor were informed that their material was obviously faulty and must be completely replaced at their cost (estimated to be £30,000).

Site investigation showed that the single-part polysulphide sealant used around the windows had not cured fully. Examination of samples indicated that the cure process had started normally but appeared to have stopped after 5–10 days. It appeared that after that time the seal had been attacked by some agent that caused the suspension of the cure mechanism, and also attacked the polysulphide polymer itself, releasing the terminal mercaptan groups and preventing any further chemical cure.

From the state of cure and the condition of the seal, it appeared that some chemical, probably a strong acid, had attacked the seal. After much detective work it was established that a few days after the windows had been sealed, the brickwork had been cleaned with a brick cleaning solution containing hydrochloric acid. Hydrochloric acid, when brought into contact with uncured one-part polysulphide will destroy the curing agent and will attack the polysulphide polymer, releasing the terminal mercaptan groups producing an obnoxious smell. Fully cured sealant is not normally affected by brick cleaning solutions provided that they are used in the correct dilution and washed off within the prescribed time.

Remedial work

The source of the smell was the body of the sealant and hence it was necessary to remove completely all the existing material to eliminate the smell and to ensure satisfactory adhesion. The bulk of the material was cut out and the remainder removed by scraping and cleaning. The joints were then resealed with fresh one-part polysulphide sealant.

The remedial work was completed in 1990 and there have been no further complaints of unpleasant odours.

7.16 Use of abseil technique

The project concerned was a large housing complex built using a prefabricated large concrete panel system. The vertical joints between panels were originally sealed with a polysulphide sealant that had given good service but was now needing repair or replacement.

The problem was primarily one of access because the buildings incorporate many protrusions such as bays or balconies which make it difficult to use large cradles or work platforms. Scaffolding is very expensive and creates security problems. Small cradles involve frequent moves with correspondingly high costs and slow progress.

A contractor proposed to use abseil techniques to overcome these difficulties but this special skill imposes its own limitations as all equipment and materials have to be carried by the operator. A single-part polyurethane sealant was chosen for the remedial work as this was available in a large flexible pack, reducing the number of packs needed, which when empty and collapsed was light, occupied little space and could easily be carried off when work was completed.

This sealant package was especially convenient to the abseil technique as it reduced the number of packages required and the empty packs were less bulky than cartridges and less of a problem to the operator (Figure 7.8).

7.17 Smelly sealant

Several large blocks of flats in Scotland were constructed with steel windows set in concrete surrounds. The windows were sealed to the surround, externally and internally, with a hot-applied acrylic sealant primarily intended for external use.

It was subsequently found that the smell of the sealant lingered inside the building for a very long time. The smell of the acrylic monomer was very unpleasant and the client would not accept the buildings on completion as

Figure 7.8 *Use of abseil technique (case history 7.16)*

they considered the smell unacceptable to their tenants. All the internal sealant had to be removed, surfaces cleaned and the joints resealed with a two-part polysulphide before the buildings were accepted.

Sealants used externally may produce unpleasant smells that are acceptable on the outside of the building where the odour is dispersed by the wind. Such odours can be unacceptable indoors or in confined spaces. Many sealants produce odours or solvent vapours until cured or dry and special consideration may be necessary for internal applications.

7.18 Staining by silicone sealants

This case concerned a large office block which was clad with a light coloured marble. The recessed butt joints between the marble panels were sealed with a neutral cure silicone sealant in conjunction with a priming system.

After a few months a stain appeared adjacent to the joints, progressing from the edges of the joints and extending into the stone producing a band some 25 mm wide.

The cause of the problem

This phenomenon is typical of migratory staining from oils or plasticizers in some sealants, certain types of natural stone being more prone than others to the effect. Marble is not necessarily more susceptible to staining than, say, granite, sandstone or Portland stone, but experience has shown that some types of marble and granite are sensitive to such staining.

Consideration and testing

The only way to predict likely problems is to carry out staining tests using the same type of stone, primer and sealant materials proposed for a particular building. Such tests should take account of the loading and environmental conditions that are anticipated in service. Experience has shown that migratory staining is more likely to occur when a sealant is subjected to compression, rather than when unstressed or in tension.

With most types of sealant there is a risk of migratory staining of natural stone. It is therefore essential that a primer is used which forms an effective barrier between the sealant and the stone. Not all primers provide such a barrier, but even if one of the correct type is chosen its application plays a very important part.

Applying sealant to the face of a joint can also lead to problems of migratory staining. This is because where natural stone has been cut, slight chipping of the corners and edges is inevitable, making it virtually impossible to ensure the deposition of an unbroken layer of primer. Further, there is a major risk of displacing the primer onto the visible face of the stone. Therefore a better option is to recess the joints.

The major problem with resealing the joints of this large office block is removal of the existing stain. The stain is inside the stone, it is probably equal to the depth of the sealant and will have already penetrated across the panel. If the source of the stain, that is the sealant, is removed, the intensity of the migrant stain can be reduced by heat, thorough washing with an appropriate solvent or by use of silicone digesters. However, the problem with the use of solvents or silicone digesters is that they may themselves leave a stain; even though the intensity of the original stain may be reduced, the area of staining may actually be increased.

Remedial treatment

There is no easy solution to this problem. The main difficulty is removal of the existing stain.

The original problem is not with the design of the cladding, or with the recessed joints. The problem is one of the susceptibility of the particular type of stone cladding to migratory staining and with the choice of the particular sealant/primer system used.

If the stain can be removed, or at least its intensity reduced to an acceptable level, it should be possible to reseal these joints successfully using an alternative sealant/primer system. Fortunately the choice of sealant materials has recently been increased with the introduction of specially formulated non-staining silicone or other sealants.

8 Lessons for new build

8.1 The importance of the joint

The primary purpose of a building is to provide or enclose an area with a controlled environment. The joints in the walls or roof of that enclosure are as important in maintaining that controlled environment as the cladding panels, the windows, the roof covering or any other part of the envelope. Joint failure means failure to maintain control and, therefore, failure to provide the building performance required. (**Section 1.2**)

Joints should not be regarded as convenient gaps for the contractor to accommodate tolerances, or as unfortunate necessities to be made as small as possible and shamefully tucked away out of site. The joint is a vital part of the building and needs to be designed, constructed and sealed in a professional manner in order that it can perform its function throughout the life of the building.

Sealant application in modern buildings is not just filling up holes with a sticky mastic. It is a critical part of the construction and needs to be treated as a critical component of the cladding, with proper provision in the work programme to allow the joints to be completed in a manner suitable to ensure that their performance is adequate for the purpose intended. A major problem is that a finished sealed joint may look perfectly sound, with defects only becoming apparent at a later date.

Joint sealing is usually regarded as a minor part of the construction process, ancillary to the real work of building. There are frequently economic and time factors affecting the sealing that may affect the quality of the installation. On many occasions the time requirements, access needs and weather conditions for proper sealant application are ignored by main and other contractors. The sealant applicator is called to site at short notice and required to complete the work in too short a time, frequently in inclement weather, and often from scaffolding that is already being dismantled. This often results in skimped preparation and poor workmanship as the applicator tries to complete the work to satisfy the contractor.

Such a case was the sealant applicator in Scotland, called to site in February to seal joints in a glazed roof that should have been ready for sealing the previous August. The applicator found that the joints were under 400 mm of snow, with the instruction that they must be completed in five days as the scaffold would be removed. What was the realistic chance of a sound installation?

8.2 The work programme

To ensure the proper performance of a sealed joint, it is necessary to provide enough time in the work programme for the application of sealants taking into account the need for access, suitable weather conditions (**Section 4.2.3**), interaction with other trades and joint protection as necessary. Early liaison with contractors is essential to ensure adequate provision for the sealing operation.

8.3 Life expectancy

One of the major parameters that should be considered in designing the building is cost-in-service; this philosophy is covered in BS 7543, the guide to Durability of Buildings and Building Elements, Products and Components. Maintenance of the weather tightness of the external envelope is an important part of that cost-in-service. Part of that maintenance is repair and resealing of joints. The frequency and the ease or difficulty of resealing will affect the maintenance cost and the cost-in-service of the building. (**Section 4.1**).

The sealant joint is subject to various physical and chemical degradation processes and the life expectancy of joints subject to movement is usually less than that of the facade. It is important, therefore, that such joints should be designed to allow resealing to be carried out without excessive difficulty. Although recessing such joints or forming joints within the cladding system may protect the sealant from direct weather, it may make resealing extremely difficult and costly or impossible. (**Section 7.8**).

Assuming the proper application of the sealant, the life of the sealed joint depends upon the:

- nature of the sealant;
- exposure of the joint to weather;
- amount and frequency of movement affecting the seal;
- other degrading or damaging influences;
- geometry of the sealant joint;
- nature of the joint faces.

Proper joint design and sealant application are primary requirements, to ensure suitable joint dimensions, preparation, cleaning, priming (**Section 5.7**), *back-up and application of sealant. Without proper joint design and sealant application, the remainder of the specification is irrelevant.*

8.3.1 Nature of the sealant

The nature of the sealant affects the life expectancy in various ways (**Section 3.2**). The elastic character of the sealant (**Table 2.1**) and the nature of the movement affecting the joint will also affect life expectancy.

Elastic sealants (**Section 2.3**) are best suited to joints affected by large and frequent movements such as occur in light and well-insulated claddings.

Elasto-plastic sealants (**Sections 1.6 and 2.3**) are best suited to joints affected by slow, cyclic, movement and to permanent deformation such as occur in heavy cladding systems and structural joints.

Plastic sealants (**Section 2.3**) are suited to small slow, cyclic, movements and permanent deformation, and have successfully sealed joints in heavy cladding systems and joints in traditional construction.

Experimental work within the RESEAL Project (**Appendix 2**) showed that it is very difficult to obtain adequate adhesion, and therefore satisfactory performance, with curing sealants in joints whose surfaces are contaminated with non-curing sealant residues. Therefore, if it is likely that a joint will be resealed in future with a curing sealant, it is inadvisable to use a non-curing sealant in the first place.

8.3.2 Exposure of the joint to weather

Exposure to weather will affect the life of the sealant. The life of the joints that are not subject to movement can be extended substantially by protecting the joint against weather. The combination of a suitable durable sealant used in a non-moving joint protected against weather can produce joints with a similar life expectancy to the cladding itself. Such joints may be very difficult to reseal but may not require resealing within the life of the cladding.

The life of joints subject to movement can be extended by protecting the joint against weather, but it is difficult to provide good protection due to the changing dimensions and geometry of the joint and the movement.

Stressing of the seal tends to accelerate degradation. As a result, the life expectancy of the sealant joint subject to movement is usually less than that of the facade and resealing will be necessary within the life of the structure. In many cases the protection of the joint against weather would make resealing difficult, if not impossible. The use of more durable modern sealants is often more profitable than protection.

8.3.3 Amount and frequency of movement affecting the seal

Joints affected by excessive movement will fail prematurely. It is important to ensure that the sealant is working within its movement accommodation capability (**Sections 2.2, 2.3 and 5.8**). *Joints rarely fail because they are too large, but they frequently fail because they are too small.*

All the movements affecting the joint should be assessed to determine the composite amplitude and nature of movement. A suitable sealant can then be selected and the minimum joint width calculated. Erection and component tolerances should then be assessed in order to determine the nominal/design size of the joints. Maximum joint size can also be assessed and checked with sealant properties and aesthetic requirements.

The design process should produce three joint widths, *nominal or design size, minimum size and maximum size.* Anything constructed outside these ranges is not acceptable, particularly if the joint widths are less than the minimum size.

8.3.4 Other degrading or damaging influences

Joints in facades are designed primarily to resist weather but some joints may be vulnerable to other influences such as vandalism, chemical or solvent spillage, etc. Such factors must be taken into account at an early stage and either the joints protected by some form of mechanical protection or the joints modified to incorporate a sealant designed to resist the effect, with the joint dimensions adjusted if necessary.

Joints in floors and areas subject to traffic require special consideration. Sealants designed to deform and accommodate the movement of components will not support heavy loads crossing the joint. The sealant will deflect under the imposed load, depressing the seal and allowing the traffic, whether wheel or foot, to impact the edges of the joint. Depression of the seal imposes severe peel stresses on the sealant/component bond and can cause adhesion failure. Impact on the edge of the joint can fracture the arris, causing spalling and destroying the sealant/component link. This problem is very common in factory floors where trolleys with small hard wheels are used to carry goods around the factory, and in areas subject to stiletto heels such as shopping malls.

The depression of the seal can be reduced by increasing the depth of seal and by reducing the joint width, by incorporating more joints. Where wide joints in floors cannot be avoided, cover plates should be used to prevent damage to the joints and joint arrises.

Reinforcement and protection of the joint arrises by incorporating metal angles to form the joint faces can also be used and are suited to decorative floors.

8.3.5 Geometry of the sealant joint

Shape

There are only two basic shapes of movement joint, the lap joint and the butt joint (**Section 1.2 and Figure 1.1**). In both cases the section of sealant is rectangular, with two opposing faces bonded to the joint faces and the other two faces free to allow the section to extend or compress, or to deform. The butt joint accommodates movement by extension and compression whereas the lap joint accommodates movement in shear, with the rectangular seal deforming to form a parallelogram. Other sealant shapes are less efficient, in that movement causes an uneven distribution of stress with consequential overstress in areas that lead to premature joint failure. It is generally very difficult to reseal lap joints.

Triangular fillets and V-joints are widely used but are only suited to very low movement joints as overstress in the narrow part of the joint frequently induces tearing, splitting or adhesion loss, resulting in early failure.

Width

The maximum and minimum width of joint are determined from the anticipated movement, the sealant performance and the joint tolerances. (**Section 2.2.1**)

Depth

The joint depth must be sufficient to accommodate an adequate depth of sealant, an adequate section of the specified back-up material, and to allow recessing of the joint if required. Sealant section depends on the function of the joint, the elastic properties of the sealant and any loading that may affect the joint (**Sections 2.2.2, 4.1.3 and 5.8, Figure 5.9**).

In general, rectangular back-up foams should be 20% wider than the joint and their thickness should be at least half the width (**Section 5.8.2**). Rod-form backing should be at least 20% wider than the joint, where joint depth is equal to the joint width. For convenience, back-up foam dimensions may be assessed for the maximum joint width. If there is insufficient depth to accommodate the backing foam, a bond-breaker tape may be a suitable alternative (**Section 5.8.3**).

Recessing

In many cases, recessing the joint is an aesthetic consideration only but, where joints are subject to large or frequent movements, recessing is a functional requirement to prevent the seal bulging forward of the cladding when subject to compression. Such bulging makes the seal vulnerable to damage, it disrupts water flow over the cladding and can cause pattern staining due to dirt collection; the seals also tend to collect dirt on their surfaces. Bulging is likely to be uneven and uncontrolled, and some bulging may be permanent due to distortion during cure. The appearance of such installations can be undesirable and recessing of the joints becomes more a practical necessity than one of an aesthetic consideration (**Section 2.4 and Figure 2.1**).

8.3.6 Nature of the joint faces and use of primers

The performance of a joint relies upon the sealant gaining adhesion, and remaining bonded, to the joint faces. Primers, being relatively low viscosity materials, assist adhesion either by penetrating the pores of a porous surface or by forming a chemical link between the surface and the high viscosity sealant. Primers may also bind and reinforce weak surface layers of certain substrates (**Sections 5.2, 5.3, 5.4, 5.7 and 6.9.4**).

Some substrates may be weakened by moisture, frost, weathering, and so on, leading to a deterioration in the bond to such materials; the correct choice of primer will often minimize such effects. Experience has shown that with proper joint design and sealant application, the use of appropriate primer materials results in sealed joints of the greatest durability.

Where painted surfaces are involved, care must be taken to ensure that the paint will remain securely bonded to the substrate and to ensure that there is no detrimental interaction between the paint and sealant, sealant cleaners or primers.

8.4 Timing

Many of the permanent deformation movements in buildings occur relatively early in the life of the building. Such movements commence with construction, and continue through the construction period and into the early life of the building. If the sealing of such joints can be delayed part of the permanent deformation will have already occurred, reducing the amount of movement to be accommodated by the seal, thus reducing the stress in the joint and the risk of failure. (**Sections 1.3, 1.4, 1.5 and 1.11**).

Delaying the sealing of structural, and other, joints to accommodate permanent deformation as late as possible in the construction programme can reduce the amount of movement to be accommodated, and hence reduces the risk of joint failure.

8.5 Resealing

It is almost certain that resealing will be necessary within the life of a building. Consideration must therefore be given to resealing at the design stage to ensure that a resealing operation can eventually be carried out efficiently and at reasonable cost when necessary, thus ensuring an economic cost-in-service for the building itself. In many cases, adequate consideration of resealing will also ensure that the initial sealing can also be completed in an efficient and reliable manner, thus reducing initial costs and reducing the risk of premature failure.

8.6 Access

Joints should be accessible to permit resealing. Operators must be able to get to both sides of the joints, without excessive difficulty or the need for expensive access equipment. The use of mechanical aids for cleaning should not be overlooked. (**Section 4.1**).

Joints should be designed to permit the original sealant to be removed and the joint faces cleaned to receive the new sealant system. Smooth joint faces are an asset in this respect whereas complex profiles, nibs or grooves, or rough uneven surfaces such as exposed aggregate, make reliable cleaning difficult – if not impossible. (**Section 5.4**)

The presence of bituminous coatings, dpc's, and so on in the jointing area can be a serious impediment to cleaning and preparation of the joint.

The joint surfaces should preferably be materials than can be cleaned with commonly used cleaning solvents and specifically suited to be cleaned with solvents appropriate to the original seal.

Where necessary, removal of the existing joint backing and replacement with new material should not be difficult. Where joint backing cannot be removed, covering with a bond-breaker tape may be necessary to ensure proper sealant performance. (**Figure 6.1**).

Appendix 1 Headings for Survey, Specification and Assessment sheets.

These pages are not intended to be proformas ready for use. Some relevant section numbers for reference in the guide are shown in italics.

Appendix 145

RESEAL BUILDING SURVEY SHEET

The Building	
Address	
Location	
Building orientation(s)	

Survey details	
Date of survey	
Time of survey	
Temperature of air	
Temperature of surfaces	

Existing seal system	
Application	S3.1.6
Depth	
Width	
Back-up	
Age of seal	
Condition of seal	

Other details	
Access limitations	S4.1.2, S4.2.2
Other requirements	

Joint details	
Surface 1	
Surface 2	
Length	
Width (± variation)	(±)mm
Depth (± variation)	(±)mm

Defects observed	
Patterns	S4.2.1
Scale of damage	
Contamination	
Cause of failure	S3.1, S3.2

Joint assessment	S1
Joint function	
Movement estimated	S1
Movement designed for	S2.2
Cause of movement	S1
Frequency of movement	

Observations
Please sketch the building. Note on patterns and features which may be important in the assessment of the contract.

RESEAL SPECIFICATION SHEET
See S4.2, S6.5

Joints	S2.2, S5.8
Min Width	
Max Width	
Min Depth	
Max Depth	
Sealant width to depth ratio	

Aesthetic considerations	
Colour	S2.5.1
Finish	
Tooling	S5.11

Back-up details	S5.8
Back-up type	
Back-up shape	
Compression required	
Minimum size	
Orientation for fitting	

Operator instructions	
Sealant type	S2.2
Primer type	S5.7
Procedure for variation of specification	
Protection of seal in cure	S5.13

Preparation required	S4.1.3, S4.2.2, S5.2, S5.3, S5.4

Removal of seal	S4.1.3, S5.5

Storage of materials	S6.7

Standard required	

Temperature required for application and storage	

Final preparation required	S5.6

Waste disposal	

Limitations	
Weather	S5.13
Environment	
Chemical protection required	
COSHH considerations	

Quality control procedures	

Inspection procedures	

Standards	

RESEAL ASSESSMENT SHEET

Site access and services	

Movement requirements	
M A F Amplitude Frequency	

Primer requirements	
Application method Primer open time	

Compatibility of the sealant system	

Cleaning requirements (Hot-applied or Mixing)	

Sealants' limitations	
Weather - Temperature - Humidity Cure requirements Live-time	

Materials/process limitations	
Volatile materials Fumes Smells	

Service conditions	
Acid Alkali Solvents Fuels Oils Bacteriological attack Mechanical damage	

Material options	

Appendix 2 Summary of the work completed during the RESEAL project.

RESEAL was a UK research project which ran between 1989 and 1992 under the SERC–DoE LINK Programme on Construction Maintenance and Refurbishment (CMR). The purpose of RESEAL was to study the performance of sealant materials applied to contaminated substrates and hence provide some objective guidelines for the resealing of building structures.

Oxford Brookes University was the lead partner of the project which was part-funded by a consortium of industrial companies, which included: Morton International Limited, Sika Limited, Adshead Ratcliffe & Co Limited, Fosroc-Expandite Limited, Evode Limited and Taywood Engineering Limited, who were also the main industrial participants. This appendix represents a brief summary of the experimental work conducted within the RESEAL project.

A2.1 Background

Resealing of joints in a typical building structure needs to take place a number of times within the life-time of the building. A well-sealed joint has a life expectancy somewhat less than that of a typical building, perhaps 25 years compared to 60 years or more. A poorly sealed joint has a life expectancy very much less than that of a typical building, frequently less than 10 years. Thus resealing buildings is a widespread activity but often gives cause for concern. The resealing process is more difficult than sealing joints for the first time and is complicated by a number of factors. Among the most significant are the removal of old sealant and the likely presence of residues left in the joint. This aspect of gaining adhesion to contaminated surfaces represented the focus for the experimental activity within the RESEAL project.

A2.2 Project approach

The RESEAL project was able to simulate some typical 'on-site' conditions through paying attention to realistic details. Laboratory simulation of typical surface conditions was central to a scientific study of the adhesion of sealants to contaminated substrates. This goal was achieved satisfactorily by fostering a close relationship between the project participants who represented different sectors of the construction industry.

In the early stages of the RESEAL project a survey of resealing activity in the UK was conducted to provide a snapshot of activity in 1990. The survey highlighted a number of important points:

- 55% of the joints failed within less than 10 years service; only 15% had lasted for more than 20 years.
- Adhesion failures predominate in real structures.
- Joints in cladding and window frame-to-surround were the most commonly resealed joints.
- Concrete and aluminium substrates were commonly encountered.
- Many joints were not constructed as designed.
- Joints whose surfaces were primed, and for which sealant material was correctly specified and applied, lasted well.
- Incorrect sealant system specification and poor workmanship in installation were responsible for premature failure.
- Curing sealants, especially one-part products, were commonly used in resealing. Of these the widespread use of silicone materials was apparent – an important factor in considering resealing in the future.
- Methods for the removal of existing sealant residues include cutting and scraping, abrasion and solvent cleaning. However, a film of existing residue is often left behind, and it is to this surface that the resealing system has to adhere.

These findings mirrored those of a survey, similar in scope but with a sample size five times larger, carried out in Japan in 1984. ('Improvement system of waterproofing by sealants in Japan,' Japan Sealant Industry Association, Proceedings of Kyoto International Conference, May 1992). Among other things, the findings from the survey were used to select the sealant and substrate materials for the experimental work programmes. It also highlighted the high incidence of adhesion failures in real life joints.

A large amount of preliminary experimental work was used as a learning curve to identify the very significant number of variables and problems likely to arise in later stages of the experimental programme. Two primary threads of activity were then developed through working groups and consultants to guide the development of the project:

1. Scientific analysis of sealed and resealed joint performance. This included aspects of curing, testing, surface analysis and material characterization. In the latter stages of the project, adhesion to a wide variety of contaminated substrate surfaces (i.e. resealing) was investigated involving six test laboratories.
2. Development of this document, *Resealing of Buildings : A Guide to Good Practice*, based upon the experience of the industrial participants and backed up by the results of the experimental work. Independent authority and scientific support has been given to the *Guide* from the RESEAL project work summarized in this appendix.

A2.3 Experimental work

A2.3.1 Overview

An enormous programme of experimental work was conducted during this research project. Nearly 4000 sealant joints were tested (tensile adhesion, peel and lap shear configurations), to investigate a large number of important experimental variables. This was necessary in order to realize the ambitious objectives of the project, the principal one being the assessment of resealed joint performance.

A large number of experimental parameters were addressed – selection of materials to be used (sealants, substrates, primers); sourcing of substrate materials of consistent quality; selection of meaningful test methods; investigating the importance of curing test joints and identifying what regime to use; use of spacers and spacer materials in joint fabrication; identification of key experimental variables to be investigated, e.g. which contaminants to use for the contaminated work programme; choosing the appropriate scientific analysis techniques; and finally, establishing an achievable and coherent programme of experimental work incorporating all of the above factors. This process took a long time but its contribution in making the project a success should not be underestimated as it provided a clear direction for the experimental work to follow.

The experimental work was split into two stages, the first being concerned with examining sealant performance on uncontaminated surfaces. The second, larger, stage of the project was concerned with evaluating sealant performance on contaminated surfaces, i.e. under resealing conditions.

Sealants evaluated
oxime-cured silicone (sil o)
benzamide-cured silicone (sil b)
1-part polyurethane (1-PU)
2-part polyurethane (2-PU)
1-part polysulphide (1-PS)
2-part polysulphide (2-PS)
solvent acrylic (acr)

Substrates evaluated
anodized aluminium
powder coated aluminium
concrete
GRP

Substrate surface contamination used – both as 'thin-layer' (smear) and 'thick-layer' (0.5 mm) contaminants:

benzamide-cured silicone (sil b)   }
1-part polyurethane (1-PU)   }'curing' contaminant sealant residues
2-part polysulphide (2-PS)   }

acrylic (acr)   }
butyl rubber   }'non-curing' contaminant sealant
bitumen   }residues
GP mastic   }

A2.3.2 Joint fabrication and curing

Joint fabrication and curing are discussed together because the way a joint is assembled has a direct relationship on how the sealant material in it cures. In the initial work programme on uncontaminated substrates the joints were assembled using two polypropylene spacers to control joint dimensions. This was changed to one spacer for the second-stage work programme on contaminated substrates, as this allows faster curing for the one-part sealants and is more representative of real-life joint configurations. The whole aspect of appreciating sealant curing was addressed during this project and is an aspect of sealant performance which has tended to be ignored in previous published research.

The significance of the curing regime used emerged from this research. Standard curing regimes use controlled conditions of 40°C/95% RH for one-part sealants and 23°C/65% RH for two-part sealants. The 40°C/95% RH condition is in effect an accelerated curing regime which gave rise to some anomalous results and was part of the reason why outside curing was adopted for the bulk of the experimental work.

A2.3.3 Mechanical testing

The mechanical testing employed in this programme was quasi-static in nature – tensile adhesion, peel and lap shear tests. Sealed joints in real buildings experience cyclic strains and therefore cyclic stresses. However, the mechanical testing used in the experimental work provided very useful information about sealant performance in reseal situations.

For the main work programme in assessing sealant performance on contaminated substrates, the tensile adhesion test was used exclusively. This was adopted on the evidence from the initial uncontaminated work programme which showed that the tensile adhesion test was more reliable and consistent compared to the peel test. The key properties to record from tensile adhesion testing are the failure mode and extension at peak load exhibited by the joint, together with modulus. These parameters proved to be quite revealing in assessing a sealant's adhesion to contaminated substrates.

It was observed during the course of the mechanical testing that the conventional sealant tests were not ideal for assessing the absolute adhesion potential of a sealant. Some initial investigations were therefore conducted within the RESEAL project using some non-standard tests, specifically 'thin bondline' lap shear tests.

A2.3.4 Surface analysis

The use of scientific surface analysis techniques to help interpret the mechanical testing results provided a novel element in the research, and should help to raise the technological profile of sealant science. The techniques used were X-ray photoelectron spectroscopy (XPS); Fourier transform infra-red spectroscopy (FTIR); contact angle/surface energy measurements; electron and optical microscopy and surface profilometry. The main purpose in using these techniques was to provide information on the various surfaces investigated in terms of their surface chemistry, morphology and energy.

Surface energy measurements, from contact angle studies, are extremely useful in investigating a sealant's adhesion to a given substrate. Such measurements can also help to predict the potential adhesion of a sealant on a particular substrate. The high correlation shown from the predicted performance from surface energy measurements and the actual results from mechanical testing of contaminated sealant joints is an example of this.

XPS studies showed that what was considered a good cleaning operation, in commercial terms, still leaves significant residues on the substrate surface.

The development of an on-site test for contact angle/surface energy measurements could very usefully help to predict the level of adhesion of a given sealant to a particular surface (provided that the surface energy characteristics of the sealant are also measured).

A2.3.5 State-of-cure studies

An important aspect of the research was the awareness shown towards the importance of state-of-cure of sealants, and a knowledge of the level of cure of sealants when they are tested. In most other published research work on sealants the question of 'cure' has almost been ignored, with statements or observations such as '... full cure was assumed after a period of four weeks ... ', being typical. The RESEAL project work on state-of-cure has shown that such statements are virtually meaningless and incorrect.

The relatively novel use of Dynamic Mechanical Thermal Analysis (DMTA) studies of glass transition temperature (T_g) was employed along with the measurement of modulus. Both properties were measured against cure time (up to 16-weeks) using various curing regimes (40°C/95% RH, 23°C/65% RH, and outside curing). This allowed Oxford Brookes University to appreciate what the level of cure was when the joints were tested after 6 weeks cure. In a long-term research project when tests are conducted at different time periods, it is important for comparative purposes to achieve similar levels of cure in joints made with different sealants and to appreciate what that level of cure is.

A2.3.6 Effect of QUV ageing

Accelerated ageing using a Q-Panel Company's cyclic ultraviolet weathering (QUV) tester seeks to combine the effects of UV, heat and moisture degradation. It was found that 2000 hours of QUV-340A ageing did not have a large effect on the failure modes of sealed joints on various different surfaces. This could indicate that 12 weeks weathering under the conditions used was insufficient to test adequately the long-term potential of a sealed joint. An important missing factor in this accelerated ageing regime was an element of mechanical joint strain. Various published studies have demonstrated the importance of mechanical strain in durability studies and of the synergistic effect that combinations of cyclic strain/UV/heat/moisture have in reducing the life of sealed joints. In future durability studies of sealed joints the effect of cyclic mechanical strain should be investigated.

It was also found that a number of sealant/contaminated substrate combinations actually exhibited improved performance after QUV ageing.

QUV ageing does promote extra cure reactions and it is possible that some of these could occur close to the substrate/sealant interface, helping to improve adhesion.

A2.4 Synopsis of results from the study of sealant performance on contaminated substrates (resealing)

The key observations from this large-scale evaluation of sealant performance on various contaminated substrates are given below. It should be remembered that the observations relate to various specific sealant/substrate combinations (A2.3.1).

1. Thick-layer (0.5 mm) contamination resulted in a significant reduction in performance, particularly joint extension, for all sealants tested. This type of contamination should be avoided at all costs in commercial resealing. Surface energy measurements showed that a thick sealant residue possesses a lower energy than either a clean or slightly contaminated substrate surface. Good wetting, and therefore adhesion, cannot be expected.
2. Priming a thick-layer-contaminated substrate surface did not bring about a significant improvement in performance for any of the sealants. There remains a need to develop primers for contaminated substrate surfaces.
3. Thin-layer (smear) contamination reduced the performance of sealants compared to their performance on clean substrates. However, the extension at peak load and joint failure modes remained acceptable (e.g. extensions at peak load typically $\geqslant 100\%$). The long-term performance associated with this surface is unknown. The non-curing oil-based contaminants are best avoided, especially on porous surfaces.
4. A significant drop in extension by a sealant on a given surface is very often an indicator of poor adhesion to that surface.
5. It has been shown that cutting back a sealant residue as far as possible, and then solvent-cleaning (producing 'thin-layer' contamination), still leaves behind a significant sealant layer. This affects the surface chemistry and reduces the surface energy of the substrates that the contamination is on. It also reduces the 'micro-roughness' of micro-rough substrates such as anodized aluminium and concrete. These factors have the effect of reducing the adhesion of new sealants.
6. The effect of various different sealant contaminants can be anticipated with reasonable accuracy from surface energy measurements of the contaminated surfaces. That is, reasonably good adhesion is obtained on

high surface energy surfaces (polysulphide contaminants) and poor adhesion is obtained on low surface energy contaminants (e.g. GP mastic contaminants).
7. The order of adhesion, found from tensile adhesion testing, for the various sealant contaminants was:

2-PS > 1-PU > acr > butyl rubber > sil b > bitumen > GP mastic

best surface   *worst surface*
for adhesion   *for adhesion*

This order of performance compared quite well to that predicted from surface energy measurements. *Thus better adhesion is gained to curing contaminants rather than non-curing contaminants when resealing with a curing sealant.* However, silicone residues are difficult to remove; they represent low energy surfaces and are quite difficult to adhere to.
8. It was found that a given sealant adhered well to itself as a contaminant. This statement cannot be reliably extended to generic type adhering to generic type.

A2.5 Summary resealing guidelines

The main observations result from the RESEAL project, which are briefly summarized above, were incorporated into Sections 4 and 5 of *Resealing of Buildings : A guide to good practice.*

The following general conclusions can be drawn from the results of the experimental work in the context of adhesion and joint performance.

1. Optimum performance is obtained on a clean, primed surface.
2. Traces of sealant residue will almost always result in a reduction in joint extension and overall joint performance, in relation to the type and amount of residue.
3. Non-curing and silicone sealant residues, particularly on porous surfaces such as concrete, have been found to severely reduce the performance of joints resealed with curing products.
4. Thick-layer contamination ($\geqslant 0.5$ mm) should be avoided at all costs, especially if the residue is non-curing. The use of primers appropriate to the substrate material is unlikely to help the reseal.
5. A thin layer, or smear (<0.1 mm), of a curing sealant residue such as polysulphide should provide acceptable adhesion performance. The presence of a two-part polysulphide contaminant was found to be the

least detrimental to the performance of joints resealed with curing materials.
6. Silicone sealant residues should be treated with caution.
7. Identical products will probably stick to identical products, but generically similar materials will not necessarily adhere adequately to each other.

Where a layer of sealant is to be left in place, trials should be carried out to check the adhesion of the reseal system. Any effect of cleaning solvents or primers, on the material or on the bond, should be noted. Any reduction in joint extension, or increase in adhesion failures arising from the trials, should be noted. Evaluation of potential adhesion using wettability tests/surface energy measurements is recommended, combined with the use of tensile adhesion testing. The final specification should be determined in accordance with the results, with the final decision on cleaning requirements resting with the specifier.

Glossary of terms and expressions

Ablative
A fire-rated material that works by the controlled sacrifice of material over a given time period.

ABS
Acrylonitrile-butadiene-styrene.

Adhesion failure
The loss of adhesion to one or more faces of a sealed joint.

Age-hardening
The hardening of a material as a result of the combination of prolonged cure and oxidative processes under the influence of heat or light as a result of weathering.

Back-up foam
Material of a cellular nature used to support the construction of the joint, and to contain and control the spread of the sealant to form the joint.

Bond-breaker tape
A tape to which the sealant will not adhere, or to which it adheres poorly, used to prevent three-sided adhesion thus allowing the movement experienced by the joint to be uniformly distributed across the joint.

Butt joint
A joint between two faces where there is no overlap, such that the joint will be subjected to movement by compression or extension.

Closed-cell foam
A material where the bubbles form discrete chambers in the body of the material and where there is no passage from one cell to the next.

Coefficient of linear thermal expansion
The coefficient for a given material may be taken as the increase in length per unit length of material, per degree temperature rise. The units of coefficient

can be either Kelvin or degrees centigrade.

$$\text{coefficient} = \left(\frac{\text{change of length}}{\text{original length}}\right) \times \text{temperature rise}$$

Cohesive failure
The failure of the sealant by means of a tear through the body of the seal.

Compatibility
Compatibility may be defined as the ability of a material to wet out onto another; compatibility does not, however, guarantee adhesion.

Contamination
The presence of any material which will in any way interfere with the formation of the seal.

Coupling agents
Materials which have both organic and inorganic characteristics and can therefore provide a chemical connection between organic and inorganic substances. Silane and titanate coupling agents are routinely used to confer chemical bonding between organic polymers and glass, metal or siliceous surfaces.

Cracked
A defect generally in the surface of the seal whereby fissures extend across the surface. These fissures may be along the line of the seal or across the seal, the orientation being dependent on the forces that are being applied to the joint.

Creep
A continuous or semi-continuous deformation with time under constant stress at constant temperature. The nature of the strain/time relationship for a particular sealant material and temperature depends upon the stress.

Cure
The cure of a seal may be considered to be the process by which the seal reaches the expected level of performance. Strictly speaking the cure implies the chemical cross-linking of the polymer in the material. However, the industry has in the past tended to apply the term to the drying out of solvent-based materials as well.

Damp
The presence of moisture in the surface. This may be in the form of water vapour or water within the body of the substrate. It is not the presence of water on the surface.

Damp-proof course (dpc)
Layer of damp-proof material, often bituminous or pitch-polymer based, introduced into walls and around openings to prevent moisture from migrating.

Displacement of the seal
The degree to which the seal is forced out of its normal position by the expansion or contraction of the joint as it moves.

Drying oils
Oils, generally derived from natural sources where there is a degree of unsaturation. This implies the possibility for the molecules to be joined chemically by the action of catalytic or thermal means to form a polymer.

Durability
The ability of a seal to last in a given set of environmental conditions.

Elastic
A material which deforms under an applied load but recovers immediately on the removal of that load to its original shape. The time-scale for the applied load is not important.

Elasto-plastic
A material which deforms under an applied load, but which may partially recover on the removal of that load. The degree of recovery may be rapid but partial, or it may be slow. Where an applied load is left in place for a long time the sealant may stress-relax (see stress-relaxation properties).

Extrude
The process of forcing a sealant from the gun into the joint.

Failure
The point at which the seal no longer performs the function for which it was installed.

Filler/fibre board
Boards often made from wood or mineral fibre used to form joints in buildings; these are sometimes left in position on completion of the building, necessitating their isolation from the seal. These materials are also often impregnated with bituminous materials.

Fire resistance/retardance
Care must be taken when using the terms fire resistance and fire retardance in this context. Strictly the resistance implies that materials resist fire, i.e. they are unchanged by it for a given period of time; retardance means that the

passage of fire is delayed. In practice, fire resistance for the sealant industry is generally defined by the level of performance in terms of time obtained by subjecting the sealant to test in a configuration determined by the manufacturer. The current test for this is BS 476 Part 22.

Intumescent
A method of reaction to fire. The intumescent process may be considered to be the changing of the physical/chemical form of a material by a foaming-type reaction, resulting in an increase in volume of the sealant. This is generally accompanied by the formation of a surface char that is less combustible than the original material. This low density material fills the void and prevents the passage of fire. It should be noted that these chars are often both friable and fragile. Any sealant which is intumescent in character and is subjected to any form of heating beyond that normal for its environment (say, up to 40°C) should be replaced.

Joint
The occurrence of discontinuities in a building surface.

Lap joint
A joint where there is an overlap of surfaces such that there will be a shear force applied to the sealant in the gap.

Life expectancy/lifespan
The period of time that under normal conditions a sealant of a given chemical/polymer type may be expected to perform.

Mastic (oil-based)
Generally taken to be the older-style lower performance sealants based on natural oils and in some cases modified with polymer rubbers such as butyl rubbers. These were single-part materials with a relatively low life expectancy and tend to remain as a paste under a cured skin.

Modulus
The modulus referred to is the modulus of elasticity, which is a measure of the force necessary to extend a sample of a given cross-section area over a given distance. The modulus should be quoted in force per unit area for a given extension. Certain standards quote modulus parameters for materials (e.g. BS 5889) as an indication of the 'toughness of the material'. This is indicative of the force likely to be exerted on the surface of the joint. Where weak or soft surfaces are encountered this can be particularly important since a high modulus is undesirable.

Movement Accommodation Factor (MAF)
This is a method of assessing the ability of a sealant to accommodate movement. It is used in the design of a joint to ensure that there is sufficient sealant present to accommodate the likely movement in the joint. It is recommended that a good margin of error be left.

One-part/two-part/multi-part
Single-part materials are those which are supplied in a single pack and draw on elements from the environment to bring about their cure; this can be in the form of the loss of solvent, or reaction with moisture, oxygen and heat. Two-part or multi-part materials are supplied in a number of containers and are supplied in a balanced pack which must be thoroughly mixed prior to application.

Overseal
The placing of a new joint over an existing one to reseal the surface. It is important that where this is being undertaken, due consideration is given to the need for bond breakers, etc.

Plastic
A material which deforms under an applied load and does not recover when the load is removed.

Premature failure
A failure which may be considered to have occurred prior to the estimated design life of the seal. This may be due to changes in environment or other causes.

Porous/non-porous surfaces
A porous surface can be considered to be a surface through which a material, e.g. water or oil, can diffuse. Care should be taken with terms of this type since there is an infinite range, and even two parts of the same material may have significantly different levels of porosity.

Primer/surface conditioner
A material applied to make the surface more readily able to accept the sealant and to improve the level of wetting out, and hence bonding. Primers/conditioners may also exhibit the ability to displace water from a surface, or to reduce its effect. The manufacturer's advice should always be obtained with regard to the correct material for use in a given situation.

Sealant
A material designed for the sealing of gaps/discontinuities, etc in the construction industry.

Shelf-life
The period of time after manufacture when the manufacturer recommends that the material be used. This period of time is strictly dependent on the conditions of storage.

Shore A
A scale of measurement for determination of the hardness of a material. This is sometimes used as an indication of the ability of a material to withstand traffic, or its ability to accommodate movement. The scale was not devised for either of these purposes and such interpretation should be treated with caution.

Shrinkage
The reduction in volume as a result of the loss of one constituent from a sealant. This may be the loss of a solvent or the loss of a reaction by-product, e.g. acetic acid from acetoxy-cured silicones.

Silicone digesting fluids
A range of proprietary products which have been developed to break down the chemical structure of silicone polymers and to thus allow their removal from surfaces.

Slump
The movement of a sealant as a result of the effects of gravity after the sealant has been applied and tooled. This is a formulation-dependent property and therefore manufacturer's advice should be taken when considering applications to wide joints.

Solar gain
The increase in temperature in a body as a result of direct sunlight. In buildings this can be experienced in the surface of the building, for example in cladding, or in cases where there are large areas of glass this can be experienced inside the building.

Stress-relaxation properties
Some sealants exhibit stress-relaxation properties, in that when the seal is subjected to an applied load for a long time the sealant undergoes a gradual plastic deformation so that the seal is no longer under load. This process is generally quite slow, and therefore does not affect the sealants' response to daily changes, but may over a period of time reduce the effects of such things as drying out or settlement where there is a one-off or very slow movement.

Surface cohesive failure
Failures occurring at or very close to the adhesion interface but just within the body of the seal.

Surface energy
The additional energy of atoms or molecules in the surface of a solid, as opposed to the bulk, brought about by the fact that they are at a free surface. This is defined as the energy needed to create a unit area of fresh surface, expressed as energy per unit area (mJm^{-2}).

Surface tension
The higher energy state of a liquid surface, compared to the bulk, which makes it behave as if it were under tension. This tension is expressed in terms of force per unit length (mNm^{-1}). Surface energy and surface tension are dimensionally equivalent and numerically the same.

Tooling
The process of compacting the seal to ensure that there is a uniform joint, that the whole joint is filled with the sealant, and that there are no cavities.

Ullage
The space left inside the pack by the manufacturer after the addition of the components to allow for the proper mixing of the material. For example, a 1.2 litre kit may be packed into a 1.5 litre tin to enable mixing of the material.

Wetting
A measure of the ease with which a liquid will flow across a surface, penetrating the interstices, without breaking into droplets.

Information sources, publications and standards

Glass and glazing manual
Glass and Glazing Federation, 44-48 Borough High Street, London.

Guide to sealants
Association of Sealant Applicators, 1985.

The use of mastics and sealants on site
Aluminium Window Association.

Manual on good sealant application
Construction Industry Research and Information Association, SP80, 1991.

Sealants : the professionals' guide
US Sealant & Waterproofers Institute, 1984.

BRE Publications
IP 4/90 Joint sealants and primers : further studies of performance with porous surfaces.
IP 9/87 Joint primers and sealants : performance between porous cladding.
IP 8/86 Weather-proof joints in large panel systems : identification of defects.
IP 9/86 Weather-proof joints in large panel systems : remedial measures.

British Standards
BS 476 Fire testing
BS 3712 Building construction sealants, 1985.
Many parts of this document are no longer in use due to impending work at CEN/ISO level.
BS 4254 Two-part polysulphide based sealants, 1983.
BS 5212 Cold-poured joint sealants for concrete pavements, 1975, revised.
BS 5215 One-part gun-grade polysulphide based sealants, 1986.
BS 5889 One-part gun-grade silicone based sealants, 1989.
BS 6093 Code of practice for the design of joints and jointing in buildings, 1992.

BS 6213 Guide to the selection of constructional sealants, 1982.
BS 6262 Code of practice for glazing of buildings, 1982.
BS 7543 Durability of buildings and building elements, products and components, 1992.
BS 8000 : Part 7. Workmanship on building sites code of practice for glazing, 1990.
BS 8200 Code of practice for design of non load-bearing external vertical enclosures of buildings.
BS Codes of practice for brick and concrete cladding.
HMSO
Classifications of packaging and labelling regulations, 1984.
HSE Guidance note EH40 : Occupational exposure limits.
Control of pollution, Special Waste regulations, 1980.
BASA Safe handling of adhesives and sealants in industry.

ISO - International Organisation for Standardization
The majority of ISO test methods have been adopted as European Community standards (EN) through CEN (the European Standardization Committee).
ISO 2445 Joints in buildings : fundamental principles for design.
ISO 3447 Joints in buildings : general checklist of joint functions.
ISO 6927 Sealants vocabulary : 1991.
EN 26927
ISO 9048 Determination of extrudability of sealants using standardized apparatus : 1991.
EN 29048
ISO 7389 Determination of elastic recovery : 1991.
EN 27389
ISO 7390 Determination of resistance to flow : 1991.
ISO 8339 Determination of tensile properties : 1991.
EN 28339
ISO 8340 Determination of tensile properties at maintained extension : 1991.
EN 28340
ISO 8394 Determination of extrudability of one-part sealants : 1991.
EN 28394
ISO 9046 Determination of adhesion/cohesion properties at constant temperature : 1991.
EN 29046
ISO 9047 Determination of adhesion/cohesion properties at variable temperatures : 1989.

ISO 10563 Sealants for joints : Determination of change in mass and volume : 1991.

ISO 10590 Determination of adhesion/cohesion properties at maintained extension after immersion in water : 1991.

ISO 10591 Determination of adhesion/cohesion properties after immersion in water : 1991.

ISO 11431 Determination of adhesion/cohesion properties after exposure to artificial light through glass.

ISO 11432 Determination of resistance to compression.

ISO 11600 Sealants : Classification and requirements : 1993.

Index

Abrasion, 43, 55, 70, 73, 75–78
Abseil, 42, 133
  see also Access
Access, 37–43, 61, 143
Acrylic sealants:
  ageing, 33
  properties, 17, 20
  removal, 77
Adhesion:
  failure, 29–30, 119, 149
  mechanisms, 61–65
  performance, 66–67, 76–80, 119, 154–155
  tests, 58, 108, 112, 152
  three-sided, 2, 18, 82
  to sealant residues, 43, 54, 66, 67, 76, 77, 154–156
Adsorption, 63, 64
Ageing, 33
Aluminium surfaces, 8, 115–117
  anodised, 74
  powder-coated, 64, 74–75
Application life, sealants, 90–91

Back-up materials, 18, 48–49, 82–86, 108, 109
Bandage joints, 45
Bitumen sealants, removal, 76, 129
Blistering, 23, 49, 122, 130
Bonding, 62, 64
Bond-breaker tape, 18, 86–87, 109–110
Brick surfaces, 5, 7, 73
Butyl rubber sealants:
  ageing, 33
  properties, 17
  removal, 76–77

Chemical resistance, 101
Cherry pickers, 41–42, 61
Cladding panels, 7–11, 19, 114, 115, 117, 126, 130, 134
Cleaning of joint surfaces, 43–44, 61, 65–66, 68–79, 108
Closed-cell foams, 83–84
  see also Back-up materials
Coefficient of linear thermal expansion, 7–8
Cohesive failure, 31–32
Colour, 7–10, 21, 22, 130
Compatibility, 44, 48, 54, 67
Concrete surfaces, 6–7, 63, 66, 73
Contact angle, 62–63, 152–153
Contamination, 62, 66–67, 68, 70, 76–78, 151, 154
Coupling agents, 64
  see also Primers
Cost:
  access, 37–39, 53
  -in-service, 37, 40, 53, 138
  labour, 38
  materials, 38
  of resealing, 37–39
Creep, 6, 54
Cure, 87–88, 100–101, 151, 153
Curtain walling, 19, 120, 124–126
Cyclic joint movement, 8–11, 115

Damp surfaces, 47, 80
Displacement of seal, 129
Drying rate, 5
Durability, 17, 21, 33, 138, 142, 148

Elastic sealants, 11, 17, 18, 54, 82, 139

168 Index

Elastoplastic sealants, 11, 20, 54, 82, 139
Epoxy sealants, 17, 21
Extruding, 93

Failure of sealed joints:
   causes, 15, 26, 50–52, 149
   types, 29–34
Fibre board, 49, 84
Fillet seals, 98, 124
Fire resistance, 23
Foam gaskets, 83
   see also Back-up Materials; Closed-cell foams; Polyethylene

Gaskets, 61, 120–122
Glass, 7, 62
GRP cladding panels, 7–8
Gunning, 91–94

Health and safety, 106

Insulation, 7–10, 52, 114–116
Interfacial contact, 62
Intumescent, 24

Joints:
   butt, 2
   cleaning, 43, 54, 65–75, 108, 126
   construction, 1
   depth, 4, 18, 81–82, 107, 108, 141
   design, 1, 16, 104, 141
   fillet, 45, 98, 124, 141
   movement, 1–14, 141
   narrow, 107–140
   preparation, 65, 79
   purpose, 1
   recessed, 19, 22, 142
   resealing, 37–49, 61–101, 148, 154–156
   shear, 1–2
   spacing, 51–52
   types, 1–2, 16, 141
   width, 16, 82, 104, 107, 108, 140, 141
Joint movement:
   causes, 1–12
   cyclic, 6–11
   measurement, 12–13
   sealant selection, 18–20
   types, 1–2, 141
   see also Movement accommodation

Laitance, 63
Lap joint, 1–2, 141
Life expectancy, 17, 21, 33, 35–36, 138
Loading, 11

Maintenance of sealed joints, 138
Masking, 98–99
Mastics, 33
Mechanical interlocking, 64
Metal surfaces, 62, 64, 74
Microporous finishes, 75
Modulus, 20–21, 45, 67
Moisture:
   effect of adhesion, 47, 56, 58–59, 80
   effect on movement of materials, 5–6
   effect on sealant material cure, 56
Movement accommodation, 16, 21, 53–54, 140
MAF (movement accommodation factor), 16, 17, 18, 39, 54, 82

New build, 137
Non-curing sealants:
   ageing, 33-34
   removal, 76–77

Oil-based sealants (mastics):
   ageing, 33–34
   properties, 17, 28
   removal, 76–77
Operator:
   technique, 105
   workmanship, 53, 106–107, 126, 149
Oversealing, 45–46, 120–122

Painted surfaces, 75, 142
Pattern effects, 50–52
Plastic sealants, 11, 17, 20, 54, 82, 139
Plastic surfaces, 7–8
   adhesion, 62, 64
Polyethylene foam, 18, 49, 83, 84, 109
   see also Back-up materials

Polysulphide sealants:
  ageing, 35
  properties, 17, 28
  removal, 77–78
Polyurethane sealants:
  ageing, 36
  properties, 17, 28
  removal, 77–78
Pouring grade sealants, 91
Powder-coated metals, 64, 74–75
  see also Aluminium surfaces
Porous surfaces, 64, 73–74, 142, 155
Primers, 47, 62–66, 79, 98, 108, 109, 142, 149, 154, 155
  see also Coupling agents

Resealing joints:
  access, 37–43, 54–55, 61, 143
  cleaning techniques, 43–47, 54, 61, 65, 68–79, 108
  cost of resealing, 37–39
  inspection of failed joints, 26–28
  preparation, 79, 109
  priming, 47, 62–66, 79–81, 98, 109, 142, 154, 155
  removal of failed sealant, 43–45, 66, 76–78
  specification, 104, 106
Reseal project, 148–156

Scaffolding, 41, 61
Sealant:
  appearance/colour, 22
  application, 91, 105, 108, 110
  application life, 87, 90–91
  classification, 17–20
  discolouration, 124, 127, 130
  hidden, 124
  identification, 28–29
  removal, 43–45, 68–78
  selection criteria, 20, 49–57, 103
Sealant materials:
  curing, 56, 87–91, 100–101, 153
  generic types, 17–20
  mixing of two-part materials, 78, 87–91

multi-part, 87–88
shelf-life, 105
stress relaxation, 20
viscosity, 55, 89, 91, 142
see also Acrylic; Bitumen; Butyl rubber; Elastic; Elastoplastic; Epoxy; Oil-based; Plastic; Polysulphide; Polyurethane; Pouring grade; Silicone; Temperature
Seepage, 25, 117
Settlement movement, 4
Shrinkage of substrates, 5, 13, 53–54
Silicone digesting fluids, 25, 44
Silicone sealants:
  ageing, 36
  properties, 17, 28
  removal, 78
Site:
  practice, 106–107
  trials, 57–60, 104
Solvents, 71–72
Staining, 24, 117–119
Steel surfaces, 7, 74
Stone, 7–8, 73
Surface:
  chemistry, 63, 152, 154
  cohesive failure, 31, 33, 34
  energy, 62, 152–155
  morphology, 63–64, 152, 154
  tension, 62–63

Tear resistance, 19, 33, 35, 36
Temperature, 6–11, 55–57, 103
Thermal expansion, 6–11, 51–53, 114–116
Timber surfaces, 5–8, 75, 122
Tooling, 95–100, 108, 110, 111

Ullage, 88

van der Waals' forces, 64
Vandalism, 21, 140
Vibration, 12

Weather resistance, 33–36, 103, 138–139
Wetting, 62–63, 154
Wind, 11–12